Dynamics of
Marine Fish Populations

Brian J. Rothschild

Harvard University Press
Cambridge, Massachusetts, and London, England   1986

This book is printed on acid-free paper, and its binding materials
have been chosen for strength and durability.

Library of Congress Cataloging-in-Publication Data

Rothschild, Brian J., 1934–
  Dynamics of marine fish populations.

  Bibliography: p.
  Includes index.
  1. Marine fishes. 2. Fish populations. I. Title.
QL620.R68   1986   597.05′2636   86-9877
ISBN 0-674-21879-5 (alk. paper)

For Elise and Jessica

Preface

Our perceptions shape the substance and form of the boundary between the unknown and what is thought to be known. The illumination of this boundary is the essence of science. In this book I examine the still-obscure boundary in one small field of inquiry — the causes of change in the numerical abundance of fish populations.

I began this work in Rome in 1978, when I attempted to identify constraints hindering development of a predictive understanding of variability in fish-population dynamics. The most important constraints seemed to be that only a few components of the entire problem had been studied and that continuing research seemed to focus ever more intensively on those few components. Increases in research effort did not result in commensurate increases in knowledge. Because comprehensive analyses did not exist, research outside the focus tended to be haphazard in direction and thus did not make the most efficient contribution to the overall goal of understanding fish-stock variability.

As a remedy, I prescribed a systematic analysis of the entire problem and suggested a structure for further investigation. This structure involved examining questions regarding the interrelationships in a simple configuration of biological variables relevant to fish-population dynamics — namely, egg production, nutrition of larval fish, and predation on fish larvae — and the simplest configuration of physical variables thought to influence those biological variables — temperature, light, and motion. Questions on the nature of the connections between these variables (for example, egg production and temperature, egg production and light) constituted what I called the cardinal questions of fish-population dynamics (see Rothschild, 1981).

I pointed out that the interactions were so poorly understood that it was not possible to identify which particular relationship, or set of relationships, was more important than any other. Moreover, there was little hope of gaining insight into these issues merely by observing the historical recruitment record and some corresponding time series of environmental variables.

My work on this subject lay fallow for several years until I was asked by James J. McCarthy, curator of the Museum of Comparative Zoology at Harvard University, to present the Columbus O'Donnell Iselin II Lecture. This opportunity stimulated me to reconsider my manuscript and to expand it into the material recorded in this book. I am indebted to McCarthy and his colleague Allan R. Robinson for encouraging me to develop and articulate these thoughts, which otherwise would have remained inchoate.

There is not much that we can learn about fish-stock variability from the study of fish. What we need to do is specify or identify the fish's *environment*, a most difficult task. Of course, we can be certain that fish-population dynamics are influenced by the dynamics of their associated biota; that these biota are affected by the physics and chemistry of the sea; and that the physics and chemistry of the sea are in turn driven by weather and climate. Further, we can characterize the sea in virtually the entire spectrum of time-and-space scales in terms of its physics, its chemistry, and its biota.

Little is known about these relationships beyond the simple fact that they exist. The particular variables among the infinitude of environmental variables that cause increases or decreases in abundance of one or several stocks of fish cannot be identified. It may be that time sequences of the most important variables have not even been measured because they have not been identified or are thought to be unimportant. Thus the problem of predicting fish-population variability does not at present rest on our understanding of the behavior and physiology of any particular group of fish. Rather, it rests on our ability to identify, define, and measure the fish's ambient environment.

The complexity of such a task is rooted in the virtually infinite dimensionality of the marine ecosystem. In systems of very high dimensionality, the distinction between truth and falsity is often blurred. What seems patent under one set of circumstances may seem false and contradictory under similar but different circumstances. While systems of very high dimensionality are hard to describe, assertions regarding their form are even more difficult to test or verify.

Most approaches to the problem of fish-stock variability have been descriptive and have ignored the high dimensionality of the ecosystem. A typical approach generally involves (a) selecting some time sequence of annual fish-stock abundance, (b) selecting some time sequence of abiotic or biotic environmental variable(s), and (c) analyzing the relationship between sequences. If a statistically promising relationship is not found, then the process is repeated. With such a procedure it is not surprising that many correlations are found. But it is most difficult to distinguish between relations that are causal or predictive and relations that are simply correlative, because typically there is no a priori theoretical justification for the choice or selection of independent variables, for fluctuations or changes in the nature of independent variables, or for the way in which the high dimensionality of the problem has been accounted for.

The approach in this book is to work toward a theory that will explain variability in fish-stock abundance. Any such theory has two components: one that explains causality in terms of the status quo and another that explains the evolution of the causal mechanisms. The first component is addressed here as a prerequisite to future study of the second. The focus of the presentation is generally limited to species of marine fish harvested by large commercial fisheries. There are several reasons for such a focus. First, attention to commercially harvested fish permits the utilization of numerous data and studies on fish-population dynamics. Second, the study of harvested fish populations provides otherwise unobtainable insights into the responses of population to measured perturbation, especially those that enhance population growth in numerically depressed populations and dampen population growth in relatively abundant populations. Third, most commercially harvested species have a primitive form of reproduction in the sense that they are highly fecund and exert no parental care (one example in the text, however, involves the pink salmon, which spawns in nests or "redds"). This feature permits my study to serve as a point of departure for exploring the dynamics of many species of fish that exhibit some form of parental care, which may include nest-building behaviors in marine, anadromous, and freshwater fishes. The theory I elaborate takes into account the dynamic behavior of harvested fish populations to rephrase the notions of density-dependent, density-independent population regulation and to explain population variability in terms of a population-dynamics process that accounts for *both* population stability and fluctuation. I suggest that stability is induced through an interaction between density-enhancing and density-dampening mechanisms; that efficient operation of the interaction depends on the fidelity of population-abundance signals transmitted to the population from the environment and vice versa; and that fluctuation is due either to unusual magnitudes of environmental signals or to associated noise.

My study attempts to elucidate this process, mainly by drawing on examples from various populations. Thus I have refrained from detailed comment on the dynamic behavior of any single stock and from a comprehensive review of the literature, although where possible I refer to monographic treatments that the reader may use to gain access to the literature. (For general background on the subject and a more complete entrance into the literature, see David Cushing's *Fisheries Biology*, 1981, and *Climate and Fisheries*, 1982.)

The first four chapters of this volume provide a background on the nature of fish-stock variability, examine the importance of understanding variations in fish-stock abundance, and establish a structure for subsequent analysis. Chapters 5, 6, and 7 consider the building blocks of the fish-stock variability problem, including recruitment-stock theory, the production of eggs, and the life and death of fish larvae. The last chapter brings together findings from the antecedent chapters and develops a foundation for a theory on fish-stock variability.

The book is intended for a diverse audience: fishery biologists and managers will be interested in the analysis of fish-population variability; ecologists will find a discussion of the population dynamics of a major group of animals barely considered in ecological texts; oceanographers may want to extend the population-dynamics perspective on ocean productivity from fish stocks to other populations in the sea; and population biologists and geneticists may want to think about the evolution of the population-dynamics-process model. Most of all, I hope the book will be interesting to readers who want to think about population variability and its biological implications.

Whatever may be original in my discourse results in some measure from my association with colleagues and friends. I have particularly benefited from discussions with Dayton Alverson, Lee Anderson, Vaughn Anthony, Philip Appleyard, Richard Barkley, Izadore Barrett, John Botzum, Garry Brewer, Susan Brunenmeister, William Burke, Douglas Chapman, W. M. Chapman, Colin Clark, Daniel Cohen, Bruce Collette, James Crutchfield, David Cushing, Harry Everhart, William Fox, Jr., David Garrod, Marvin Grosslein, John Gulland, Sigeiti Hayashi, Edward Houde, John Hunter, J. D. Isaacs, Rodney Jones, James Joseph, Hiroshi Kasahara, Peter Larkin, Reuben Lasker, William Lund, John Magnuson, John Marr, Donald McKernan, Peter Miyake, Hiroshi Nakamura, Tamio Otsu, Basil Parrish, Gerald Paulik, Randall Peterman, Edward Raney, Henry Regier, William Ricker, Douglas Robson, Olegario Rodriguez-Martin, Milner Schaefer, Gunter Seckel, Elton Sette, Kenneth Sherman,

Richard Shomura, Michael Sissenwine, Roland Smith, John Steele, Akira Suda, Richard Van Cleve, Fred Ward, Steven Wertheimer, James Westman, Norman Wilimovsky, John Woods, and Warren Wooster.

I am grateful to David Cushing, Neils Daan, Edward Houde, and Daniel Ware, who kindly read drafts of the manuscript; their suggestions materially improved its quality. I appreciate Howard Boyer's considerable editorial advice as well as that of Jacqueline Dormitzer. I wish also to thank Ian Morris for freeing my time to devote to this study and the National Oceanic and Atmospheric Administration, especially Joseph Angelovic of the National Marine Fisheries Service, for partial support. Much of the manuscript was written while visiting the Institut Français de Recherche pour l'Exploitation de la Mer in Nantes and the Institut für Meerskunde at the University of Kiel, and I am grateful to my hosts, J.-P. Troadec and John Woods, for many stimulating conversations.

As authors know, insights that seem better than any that have yet been conceived are shadowed by the drudgery of writing them over and over again. Pamela Blancato eased this burden by cheerfully and swiftly typing the many iterations of the manuscript. The difficult task of preparing virtually all of the many figures was undertaken by Susan Rothschild, whom I thank for this help, but even more for the encouragement that she provided.

Contents

Dynamics of Marine Fish Populations

1

Variability in Fish-Stock Abundance

The need to manage or control animal populations that either harm or enhance human welfare motivates the study of the temporal and spatial dynamics of animal populations. Interest in population dynamics has roots in ancient hunting, fishing, and herding practices; concern for human demography; and occurrences of the animal-related plagues and pestilences recorded since biblical times.

Prehistoric hunting and fishing have evolved into modern commercial, recreational, or subsistence hunting-and-fishing activities of considerable diversity. The effects or apparent effects of hunting and fishing on wild populations are well documented; while some populations have sustained hunting or fishing for centuries, others have dwindled to a low level of abundance or become extinct.

In addition to the direct effects of hunting and fishing, the dynamics of wild populations have been affected in many instances by extensive modification of the primeval habitat. Some of the modifications are obvious, but others are subtle. For example, changes induced by forestry, agriculture, and urbanization are physically evident in much of the world's countryside. In contrast, the *effects* of increased loadings of toxic and eutrophicating substances on the dynamics of wild populations are generally subtle and more difficult to identify. This is because these effects — that is, changes in population dynamics that can be directly ascribed to anthropogenic factors — cannot easily be observed, except in extreme cases. For example, the effects of pollution are obvious when fish are killed directly by high concentrations of toxic substances. Such circumstances, however, are quite unusual relative to situations where organisms are exposed to the entire ambient spectrum of toxic and eutrophicating substances in concentrations that are not obviously lethal but are supposedly harmful in some sense. These ''sublethal effects'' are measured in terms of an increased body burden of toxicants or an increased incidence of various lesions. However, the effect of sublethal concentrations upon growth, mortality, and reproduction is not well known. In other words, the way in which pollution affects the in situ population dynamics of fishes is not known and often not considered. Despite widespread concern regarding introduction of toxic and eutrophicating substances into the environment, their impact on animal-population dynamics is difficult to recognize and poorly understood.

Direct and indirect human influence on animal populations is also reflected in the influence that animals have on society. The harvests of wild animals generate both benefits and the real and opportunity costs of overexploitation and underexploitation.* In addition, examples of animal-borne pestilence and plague are abundant (rats and bubonic plague; locusts and agricultural disasters; spruce budworm and gypsy moth forest decimation). Further, the disastrous effects of the introduction of nonendemic species are well known (for example, the European rabbit in Australia and the sea lamprey in the Laurentian Great Lakes).

Society's heightened environmental consciousness and the realization that human activity can affect the abundance of animal populations as well as options for environmental management provide a contemporary setting for the familiar theme of our economic and social dependence on the waxing and waning of animal populations. We need to increase greatly the economic and societal efficiency or cost effectiveness with which public-policy decision-making processes control hunting, fishing, and habitat modification. Obviously the quality of the decision-making process is strongly influenced by our understanding of the mechanisms that cause populations to increase or decrease or to advance or retreat in space and in time.

But how well do we understand these mechanisms? What are the

---

* The question of benefits and costs of resource exploitation and optimal-harvest strategy is the subject of a considerable literature. Early books (for example, Christy and Scott, 1965) stressed the dilemma of the open-access nature of fishery resources, in which the resource manager is faced with balancing the private-welfare interests of the fishermen with the social-welfare goals of management. Crutchfield and Zellner (1961) provided an early quantitative analysis of this so-called tragedy of the commons (Hardin, 1968) problem. Recent comprehensive texts include graphical (Anderson, 1977) and analytic (Clark, 1976, 1985) studies of optimal-harvest strategies. While early works on the subject implied or stressed the desirability of reaching goals in the context of narrowly defined economic theory, more recent literature recognizes that achievement of economic goals is only a part of the process, because social and political interests need to be accommodated in any fishery-management strategy (see, for example, Troadec, 1983; Rothschild, 1983a,b).

factors that cause population abundance to increase or decrease in the absence of human activity? What are the precise effects of hunting and fishing on animal populations? What are the quantitative effects of physical-habitat modification and pollution? To what extent can we develop our knowledge to allow us to forecast the abundance of animal populations?

These questions are only partially addressed in early population-dynamics studies based on Thomas Robert Malthus' notion that although populations were capable of increasing geometrically, they did not do so, as this capacity was held in check by controlling factors. This theory appeared to underlie early population-growth models (see Hutchinson, 1978: chap. 1), which placed some emphasis on the "doubling time" of various populations (Malthus calculated that the U.S. population would double in a twenty-five-year period). The doubling-time calculus seems evident in Verhulst's analytic work in population dynamics, which set the stage for the notion of the "intrinsic rate of natural increase." In human and other low-fecundity populations, the notions of doubling time and "intrinsic rate of natural increase" seem intuitively natural. But their general applicability might be challenged in high-fecundity populations such as fish, where an individual female might spawn several million eggs each year. The concentration on models oriented to low-fecundity populations may have actually constrained advances in understanding the population dynamics of high-fecundity species.

The early models stimulated analyses and empirical studies that described and explicated with no little brilliance three seemingly independent types of mechanisms that evidently prevented animal species from attaining their potential for unbridled exponential population growth: the density-dependent mechanisms described by David Lack (1954, 1966); the density-independent mechanisms described by Andrewartha and Birch (1954); and the influence of behavior on population abundance as theorized by Wynne-Edwards (1962). (For a concise introduction to the early arguments on density dependence and independence, see Lack, 1966: 281–312.)

"Modern ecology" seems to have arisen as a school of thought separate from the early, highly empirical studies of population regulation. It seems that modern ecology had its roots in more mathematical attempts to extend the basically single-species population-control notions into much more difficult, fundamental, and complex issues that relate not to the dynamic behavior of single populations but to the dynamic behavior of pairs of species (for example, Lotka, 1925, 1956; Volterra, 1926) and then to entire communities. The staggering dimensionality and complexity of community dynamics were further increased by considering the problem not only in a static-time framework but in terms of its evolution as well.

The questions voiced by modern ecologists were critically important and evocative of even more interesting and exciting questions, but progress seemed to be dampened through the virtually exclusive use of mathematical models that could only superficially address, but not really elucidate, the large and shifting dimensionality of ecosystems. The models could have been more useful had they not considered the simple mathematical notions as self-contained explanations.

In fact, the use of simple mathematics to explicate processes that are much more complex than would be even the most complicated mathematical formulations raises a fundamental philosophical issue: should a problem be viewed as a "direct problem" or as an "inverse problem" (see Silvert, 1981)? The direct approach might be thought of as specifying the behavior of a system by means of an equation and then studying the equation, wherever possible relating empirical observations to the equation. In other words, the solution of the direct problem is bounded by the nature of the initially chosen equation or system of equations. The inverse problem might be thought of as observing the behavior of the system and developing equations to explain that behavior.

Much of modern ecology seems to derive from a direct rather than inverse approach. For example, Hutchinson's (1978: 23) justification of the logistic is a direct approach:

> We choose the logistic to study because it is general and because it is realistic in that it makes simple-minded biological sense. It is an excellent base from which to set out on further, more elaborate theoretical and, we hope, more accurate investigations. It is, therefore, abundantly worthwhile to see if it can give good approximations to population growth in the world of physical rather than purely conceptual organisms.

An inverse approach is not a priori constrained by some functional form. In an inverse approach, the trajectory of animal abundance, for example, would be observed and explanations developed relative to causes of population increase or decrease. It might be quite difficult to recover the logistic equation or one of its generalizations from a time series of population abundance.

The power of an inverse approach can be seen in such fields as the study of intelligence, where the workings of the human brain have been found generally too complex to model, so instead a set of parallel, simple models constituting the field of artificial intelligence has been developed. The idea is not so much to replicate the behavior of the brain as to develop a parallel system that can organize and motivate inquiry into the work-ings of intelligence, thus acknowledging that mathematical models are ordinarily not capable of fully addressing highly complex systems.

While modern ecology has identified important issues, it seems not to have succeeded in developing a predictive understanding of the ecosys-tem, and some of its fundamental notions are under challenge and open to considerable revision (see Lewin, 1984). For example, the question of whether the population dynamics of an animal operates in an *r*-selected or a K-selected mode represents an important component of the theory of modern ecology. Stearns (1976) gives a comprehensive review of the subject, listing the characteristics of *r*-selected and K-selected species. The theory, however, is evidently not applicable to marine fish, in that marine fish do not seem to fit into his classification. The theory is now being reexamined (Boyce, 1984).

Fisheries science has had a somewhat parallel evolution. Before World War II most studies were empirical. But among these, several theoretical and synthetic studies contributed to the development of a new paradigm in fishery science (for example, Baranoff, 1918; Russell, 1931; Graham, 1935). This paradigm was evident in the works of Ricker, Beverton and Holt, Schaefer, and Cushing.* The new paradigm was a considerable advance that did much to organize and systematize thinking about mor-tality rates, growth, production, stock and recruitment, yield per recruit, and the as-yet-elusive relation between the dynamics of fish and the ocean environment. In retrospect the paradigm fell short in its insights regarding a predictive understanding of the mechanisms that affect fish-stock abundance. In fact, such an understanding has not yet arisen, as evidenced in the collection *Key Papers on Fish Populations*, edited by Cushing (1983). The first seven papers, authored by Russell, Graham, Beverton and Holt, and Schaefer and Ricker, are indeed of conceptual interest. With the exception of Cushing's paper on stock and recruitment, however, the next eight papers are of computational rather than concep-

---

* Ricker's contribution was in his paper on stock and recruitment, published in 1954, and in his *Handbook for Computations for Biological Statistics of Fish Populations*, 1958. Beverton and Holt developed the first comprehensive model of fish-population dynamics in 1957, and Schaefer, also in 1957, discovered how to transform the curve of the rate of change of biomass as a function of biomass into a relation of catch to fishing effort. Cushing has pioneered concepts concerning the biological basis for recruitment-stock theory and has contributed much to our understanding of fish-environment interactions. His works are summarized in the recent publications *Fishery Biology* and *Climate and Fisheries*. The Rus-sell, Graham, Ricker, Beverton and Holt, Schaefer, and Cushing references may all be found in *Key Papers on Fish Populations*, edited by Cushing (1983).

tual importance. (The divergence from conceptual to computational work may stem from the increased availability of computer technology in the mid-1960s. Instead of using the models that had been developed as points of departure for making new observations and creating new theory, many fisheries scientists concentrated on using computers to generalize the simple theory or to make computations that were too difficult to derive without a computer. Examples are cohort analysis, multiple-species production models, and large-scale simulations. Most of these exercises were useful and had an important impact, but possibly at the expense of conceptual development.)

The specific causes of fluctuation in fish-population abundance are supported by a long list of speculations and a short list of facts. The facts are (1) the abundance of individual fish stocks has fluctuated for centuries; (2) evident human interaction with most marine stocks in terms of physical-habitat modification, pollution, or intensive fishing is relatively recent; and (3) fish stocks have continued to fluctuate in abundance in the presence of increased toxic-chemical loading, eutrophication, habitat modification, and fishing.

Much of the public and some scientists seem to believe that all downward trends in fish-stock abundance are caused by human activity and that all upward trends are caused by natural phenomena. Such beliefs might be aligned better with those that well up among prudent concerns for environmental degradation than with those tempered by careful scientific scrutiny. It is logically difficult if not impossible to ascribe an observed fluctuation in population abundance to anthropogenic causes without accounting for or understanding natural population variability. If variation in the abundance of fish stocks can be attributed to two causes (such as "natural" variation and anthropogenic variation), then inferences regarding the influence of one cause cannot be made without some assertion regarding the other. Thus if we were to make an inference about the effect of a pollutant on the abundance of a fish stock, we would also need to make one regarding the natural variability of the stock. Because this alternative inference is not always obvious, we are left with the implicit assumption that there is no natural variation; that is, the population would be at a constant level of abundance in the absence of the pollutant—a situation that is, of course, unlikely.

While separating the anthropogenic and natural components of variation is an intellectually interesting problem, it is also of considerable practical importance; environmental- and fishery-management public-sector agencies continually make management decisions that relate to fishing, pollution, or habitat modification. The sheer number of such public-sector decisions is not generally recognized, because there are basically two kinds of decisions: (1) those that are actually made and (2) those that are not made. While the decisions that *are* made are relatively easy to enumerate, those that are not made are probably much more numerous and more difficult to categorize. Not making a decision — which is often a de facto decision — may result from administrative processes or limited knowledge. One can easily see that the decision stream in cases where the dynamics of a resource is well understood could be quite different from that in cases where the dynamics is only poorly understood. These decisions substantially affect the benefits accrued by society from fishery resources in both the short and the long term and thereby touch on the livelihoods of millions.

In principle, a better understanding of the causes of variation in fish-stock abundance should result in improved policy- and decision-making processes. For example, if an anthropogenic activity is reducing the abundance of a fish stock and the activity is considered less valuable than the fish, then the activity can be constrained. But while anthropogenic activities may cause decreases in some fish populations, they may cause increases in others. With an understanding of the mechanisms, the effect of an activity can be carefully evaluated. Finally, in some instances a decline in abundance might be thought to result from human activity, whereas it may actually result from a natural phenomenon; a curtailment of the human activity in this instance might be wasteful. Thus, much is at stake in developing a better understanding of the dynamics of fish populations.

Examples of Fish-Stock Variability

To understand the variability of fish-stock abundance, it is important to appreciate the state of present knowledge regarding natural and anthropogenic causes of fish-stock variability. A more concrete and less abstract view of the problem can be attained by surveying a few well-studied fisheries or fishery complexes. The intent of the following survey is to introduce, with a minimum of interpretation, some problems associated with fish-stock fluctuations; to provide a setting for the questions that derive from observing fluctuations in fish-stock abundance; to indicate the frustrations in trying to explain the causes of fluctuations in abundance; and to search for elements that are common among the examples. The intent might best be met by first considering the Pacific sardine and the northern anchovy, the pelagic and demersal fish of the North Sea, and the pink salmon of the North Pacific Ocean.

Sardines and Anchovies

The sardine-anchovy example seems, at least at first glance, relatively simple. The Pacific sardine produced steadily increasing catches from 1920 to the mid-1930s. From the mid-1930s to the mid-1940s, landings were sustained at several hundred thousands of tons. In the mid-1940s both the landings and the population declined sharply (Figure 1.1). After the mid-1940s the stock remained at a very low level, despite a fishing moratorium, until very recently when there was evidence of a small recovery in sardine abundance (MacCall, 1983). In contrast, after the decline of the sardine, the anchovy population was observed to increase rapidly in abundance (Figure 1.1b).

What caused the sardine population to decline? Why did it not increase after fishing was terminated? What caused the anchovy population to increase? What is the relation of the decline of sardine to the increase in anchovy? Is sardine abundance actually increasing again and, if it is, what are the mechanisms causing the increase?

The first substantive, critical review and analysis of sardine research was presented by Frances Clark and John Marr (1956). Clark argued that the decline related to a reduction in reproductive success of the entire sardine stock; that this reduction was linked to a reduction in the biomass of spawning fish; and that the latter resulted from heavy fishing. Marr countered with the view that the decline was caused by environmental factors (for example, unfavorable temperatures), but once the stock was at a low level, declines in future abundance were induced by a reduced abundance of spawning fish.

The next major review of the sardine fishery was conducted by Garth Murphy (1966: 76). In explaining the sardine decline, he concluded that the key event occurred in 1949 when

> it became evident that the parameters of the population changed appreciably beginning with the recruitment of the 1949 year class to the spawning population. The qualitative change associated with the change in parameters was the reversed dominance of the two races comprising the population. Previous to the recruitment of the 1949 year class the northern race dominated; afterwards the southern race dominated.

Murphy believed that the decline was due to fishing: "It is improbable that the populations would have declined in the absence of fishing,

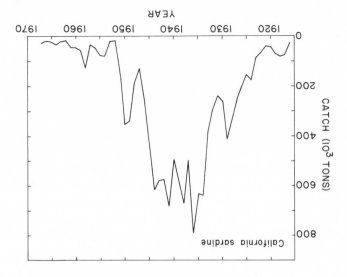

Figure 1.1a Catch of sardines along the Pacific Coast of North America. The figure shows a precipitous decline in catch starting in the 1940s. (Data from *California Cooperative Oceanic Fisheries Investigations Reports*, vols. 8 and 10; redrawn from Murphy, 1966: fig. 1.)

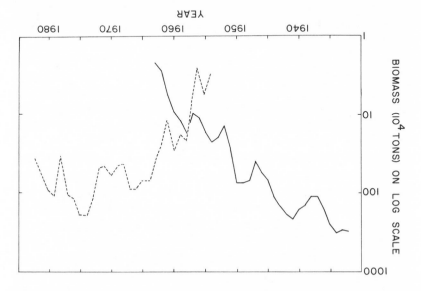

Figure 1.1b Spawning biomass of sardine (age 2+) and anchovy (log scale) off California and northern Baja California. Figure shows decline in abundance of the sardine stock *(solid line)* and increase in abundance of the anchovy stock *(dashed line)*. (Modified after MacCall, 1983: fig. 1.)

whereas the fishing rates applied to the population lowered reproduction to an extent that a decline was inevitable." He went on to suggest that the increase in anchovy was a result of the decline of sardine. Eleven years later Murphy (1977: 292) wrote that "fishing extracted sardines somewhat faster than the population could replace them, and then collapsed with the recruitment of two poor year classes in succession (1949 and 1950). Reproduction was reasonably good relative to stock size during some subsequent year, but by then the spawning stock was too small to generate a really large year class." He further observed that "the Pacific sardine generated two very large year classes (1938 and 1939), one of which (1939) was the second largest ever observed," while under heavy exploitation.

In 1981 John Radovich published "The Collapse of the California Sardine Fishery—What Have We Learned?" He reviewed much of the research on the sardine and pointed out that, among other things, Marr's correlation between sardine-year-class size and temperature may have been an artifice created by the distribution of the sardine along the coast; that acknowledgment of a far-northern race of sardine in past analyses might have resulted in alternative conclusions in Murphy's analyses; that a simulation model by Silliman (1975) purporting to demonstrate competition between the sardine and anchovy for food may have been based on misleading data; that a paper by Soutar and Isaacs (1969), based on several hundred years of depositions of sardine, anchovy, and hake scales in the anoxic sediments in the Los Angeles Basin, did not support the notion that the anchovy increased in abundance because it had gained a competitive advantage over the sardine; and that food limitations were not a factor related to the decline of the sardine (except perhaps at the postlarval stage). Radovich concluded by describing the complex of desirable properties that would be required to obtain a satisfactory model of the Pacific sardine and noted that "from the foregoing examination of only a small portion of the work which has been done on the Pacific sardine and the northern anchovy, it is apparent that most simplified generalizations are probably incorrect" (p. 132).

A more recent commentary on sardine-anchovy dynamics serves as further testimony to the difficulty of assigning causality to the fluctuations of sardine and anchovy (MacCall, 1983; Lasker and MacCall, 1983). Lasker and MacCall invoke various hypotheses concerning the decline of sardine and the rise of anchovy, including Hjort's (1914) hypothesis regarding a "critical period" in the life of very young fish; the notion that stability of the water column produces conditions favorable for anchovy survival; the notion that larval drift is important because it determines whether larvae will remain in conditions favorable for survival or will drift into unfavorable conditions; and the notion that cannibalism and changes in the geographic range of the spawning stock affect average reproductive success.

There is little basis to determine which, if any, of these hypotheses are important or to distinguish which among them are relatively more important than others. In fact MacCall (1983: 103) seems to obviate the need to consider them by observing that "the decline in abundance [of sardine] appears as a coherent trend extending over the entire thirty years of the historical fishery. There are no discontinuities which could be interpreted as adverse environmental changes." MacCall then describes rather large fluctuations in other stocks that, unlike the anchovy, have not been accorded the status of being possibly related to the change in abundance of the sardine (or of the anchovy, for that matter); for example, the Pacific mackerel declined at the same time as the sardine, and the bonito increased at the same time as the anchovy.

Thus in 1985, despite several decades of intensive study of sardines and anchovies in western North America and other clupeoids in Europe, Lasker (1985: 37) made a remarkable statement:

The answer to the question "what limits clupeoids?" seems to be "almost everything." More realistically, the question should be phrased "what limits clupeoids mostly?" Other questions follow this one: When in the life cycle does this occur? What are the interrelationships between limiting factors and between species? What can be learned from species life histories and fisheries oceanography that will allow us to predict recruitment?

North Sea Fisheries

The North Sea fishery situation, at first glance, seems more complex than that of the sardine and anchovy, but perhaps the apparently greater complexity is due to the way in which the fisheries have been described and to the relative amount of research in the two areas.

Major changes have occurred in the abundance of the North Sea stocks. For example, the abundance of the complex of herring stocks declined sharply in the decades after World War II (Burd, 1978). In contrast to the decline of herring stocks (Figure 1.2), the gadoid fishes (coalfish, haddock, cod, and whiting) of the North Sea increased substantially in abundance, roughly quadrupling their biomass during the 1960s (Figure 1.3).

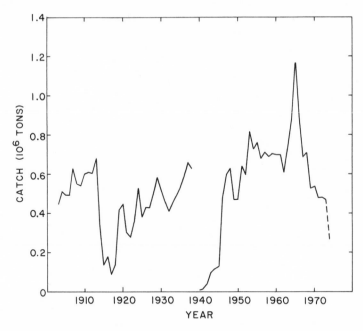

Figure 1.2 Reported herring catches, North Sea, 1903–1973. Note precipitous decline after 1970. (From Burd, 1978: fig. 128. Reprinted with permission of ICES, the International Council for the Exploration of the Sea.)

The same sorts of questions can be asked regarding the North Sea stocks as were raised about the apparently simpler Pacific sardine-anchovy example. Why did the herring collapse? Why did the gadoid fishes increase? What is the effect of fishing, and is the increase or decrease of any one species or group of species dependent on the increase or decrease of any other species?

A survey of much of the recent literature on North Sea research may be found in *North Sea Fish Stocks — Recent Changes and Their Causes*, edited by Gotthilf Hempel (1978a). Forty-eight papers were published in this volume, and later Hempel (1978b) wrote a critical review of the proceedings. In one of the papers, Holden (1978) summarized most of the available information by comparing stock abundance in the 1960s with that in the 1930s. He observed that the average catches in the North Sea had roughly doubled in the 1960s. He also noted that the increases in catches of the Norway pout, the pollack, saithe, sprat, and sand eel in the 1960s over those in the 1930s were due to postwar increases in fishing effort for these species, which were used mostly for industrial purposes rather than for food. He felt that the increase in fishing effort for these species had not exceeded their maximum sustained yield. With respect to the herring, the mackerel, and the dogfish, though, fishing effort had increased to the extent of stock depletion. But, he pointed out, fishing mortality and age at recruitment decreased since the 1930s for the plaice, cod, haddock, and whiting, and as a result the yield per recruit should have increased over 1930–1933 by factors of 1.41, 1.10, 1.25, and 1.10, respectively.

Recruitment data for both the 1960s and the 1930s were available only for the plaice, sole, haddock, and cod. Holden believed that for the plaice and cod, recruitment increases accounted for increased landings. For the sole, however, recruitment was not large enough to account for increased landings, whereas for the haddock the calculated landings appeared to

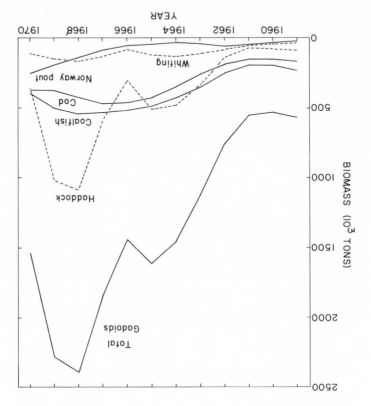

Figure 1.3 ''The gadoid outburst'': temporal increase in biomass of gadoid fish in the North Sea during the 1960s. (From Cushing, 1980: fig. 7. Reprinted with permission of ICES.)

Table 1.1 Ratios of recruitment and yield per
recruit for the 1960s and 1930s.

Fish	R_{60}/R_{30}	YR_{60}/YR_{30}
Cod	2.5	1.1
Haddock	3.8	1.25
Plaice	1.9	1.41
Sole	1.9	—
Whiting	—	1.10

Source: After Holden, 1978.

be larger than the catch, but this was evidently related to the catching of
small haddock by Danish industrial-fish fisheries. (The recruitment and
yield-per-recruit data are summarized in Table 1.1.)

Holden concluded that there is "no common feature that points to one
single factor being responsible for the high level recruitment to the cod,
plaice, sole, haddock and whiting population in the sixties . . . the only
consistent feature is that the large year-classes of sole invariably follow
winters in which sea temperatures fall markedly below average" (p. 24).
Holden discussed the correlations that he and his co-workers found
between cod and sea-surface temperatures (Holden, 1972; Dickson,
Pope, and Holden, 1974), but he also cited the view of Daan (1975)
questioning the causality of the correlation.

In a review of the symposium, Hempel (1978b: 164–165) observed,
"The discussion on the possible causes of the changes in fishing yields
over the past fifteen years might be summarized by looking at a number
of causative factors, none of which can be accepted as exclusive." Hempel identified six causal factors: (1) increased fishing intensity or availability of fish, (2) improved management of demersal stocks and overfishing of pelagic stocks, (3) increased food, producing increased growth
and earlier reproduction, (4) increased food for larvae, (5) reduced predation in early stages and adults, and (6) increased pollution.

Hempel (1978b: 165) drove home the notion of uncertainty regarding
the causes of fluctuations:

It is not possible to quantify the effects of man-made and natural factors
separately because of the complexity of interactions between the fish stocks
and the stages of their early life history. Obviously, the old fight became
pointless between those who considered fishing to be the only factor affecting each fish stock separately through growth overfishing and/or recruit-

ment overfishing and those who tried to explain almost every change in the fish production by environmental factors.

It is instructive to examine the fate of the North Sea herring and cod stocks in somewhat more detail. The stocks of herring can be divided into three groups: a northern stock, IVa; an intermediate stock, IVb; and a southern stock, IVc. They all reached a low level of abundance by the 1970s; however, their trajectories toward depletion were very different. In interpreting these trajectories it is important also to consider the corresponding trajectory of fishing mortality on each stock and to note that estimates of stock size are available for the war years, when there was no fishing and consequently no fishing mortality (Figure 1.4). We can see

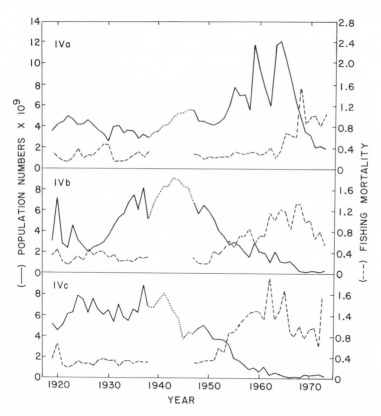

Figure 1.4 Changes in population size and fishing mortality of herring in the northern North Sea (division IVa), in the central North Sea (division IVb), and in the southern North Sea (division IVc). There was no fishing in the 1940s because of World War II. (From Burd, 1978: fig. 139. Reprinted with permission of ICES.)

that the IVa stock was at a relatively low level before World War II. Its abundance tended to increase during the war in the absence of fishing. After the war it reached a high level of abundance, owing partly to two very large year classes. Although there was a sudden increase in fishing mortality when the stock was quite large in the early 1960s, the greatest increase in fishing mortality did not occur until four years after the IVa stock began its decline.

The IVb stock was at a relatively high level just before the war; its abundance increased during the war, as did that of the IVa stock in the absence of fishing. The major difference between the wartime dynamics of the two stocks was that the increase in the IVb stock was due to a very large year class of fish produced in the early war years. The stock declined almost steadily after the war, with continuing increases in fishing effort.

The IVc stock was also at a relatively high level before the war, but in contrast to the other two stocks, it decreased in abundance with the absence of fishing during the war and continued to decline with increased fishing effort.

Andersen and Ursin (1977) linked the increase in gadoid stocks to the decline in herring stocks. Cushing (1980) examined this linkage and observed that the "gadoid outburst" might have resulted either from the "release of food" (that is, as the herring stocks declined, food that the herring would otherwise have eaten—such as plankton—became available to the gadoids) or from "relaxed predation" (that is, the possibility that the adult herring fed on small gadoids, and the decline in herring stocks reduced the mortality of postlarval cod). Cushing concluded that it is difficult to subscribe to either the release-of-food hypothesis or the relaxed-predation hypothesis and suggested a third possibility, namely, that climatic events generated the increase in gadoids independently of consequences of interspecific interaction.

It might be added that Daan (1978a: 55), after extensive examination of the increase in the cod stock, concluded that

> there is no easily detectable abiotic or biotic parameter which can be held responsible for the sudden change in the recruitment pattern . . . instead a rather complex mechanism has to be supposed that seizes upon different stages of the early life . . . it would not be unlikely that a strong correlation between recruitment and one environmental parameter or another obtained during a period of relative stability would fail to fit new observations at a different level of exploitation of the ecosystem.

A more recent analysis of the North Sea stock fluctuations is provided by Cushing (1982: 184–226), who summarizes much of the information in the North Sea symposium and considers other sources as well regarding temporal and spatial changes in the North Sea plankton, pelagic fish (herring and mackerel), and demersal fish (principally the gadoids and the flatfishes). He reemphasizes his belief that the observed changes in the biota have resulted from recent climatic change, that is, a period of climatic warming and subsequent cooling, an increase in the arctic intermediate water in the Faroe-Shetland Channel, and an increase in the number of days of westerly wind over the British Isles.

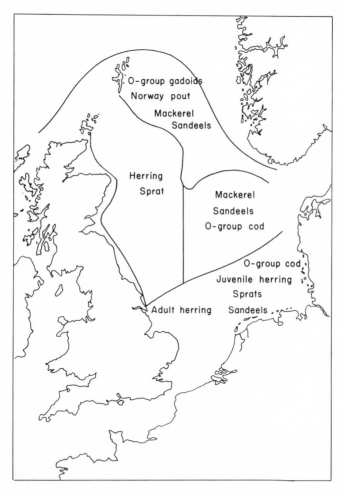

Figure 1.5 Distribution of principal pelagic feeding stages in the North Sea during the summer. (From Jones, 1983: fig. 1.)

Cushing (1982) relates these climatological and hydrographic changes to temporal changes in zooplankton and phytoplankton, which are as remarkable as the changes in fish stocks (see also Jones, 1983). Cushing asserts that the reduction in herring abundance is "linked partly to the increase in *Calanus*." The implication is that (a) herring recruitment was reduced; (b) herring abundance was reduced; and (c) the reduced herring population and the increase in *Calanus* increased the available food for herring, and these individuals grew faster, thus evidently explaining the increased recruitment in the North Sea stock in the 1960s but not quite explaining the collapse of all three stocks. With regard to the demersal stocks, Cushing attributes the fall and rise of haddock stocks to the period of cooling and then warming (see also Cushing, 1984).

Jones (1983, 1984) has also published papers on the changes in the relative abundance of various species in the North Sea. His data are from a more recent period than Holden's, but the changes are qualitively the

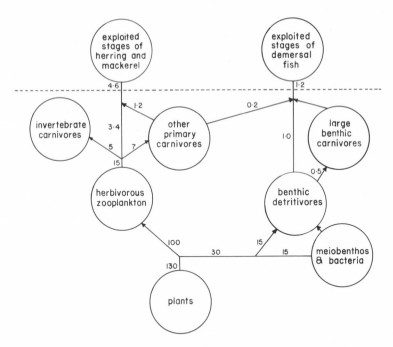

Figure 1.6 Examples of energy-flow estimates in gCm⁻²y⁻¹ in the North Sea in the early 1960s, assuming 20 percent transfer efficiency to fish through other primary carnivores and 15 percent transfer efficiency through herbivorous zooplankton. (From Jones, 1984: fig. 7a. Reprinted with permission of ICES.)

same: a decrease in pelagic stocks and an increase in demersal stocks. Jones implies that if the changes in abundance relate to interactions among the species, then these interactions should be evident in the temporal-spatial coexistence of the interacting species. As might be obvious, not all species "coexist" in the North Sea; in this regard Jones presents information, at least for the summer months, that show which species overlap and which do not (Figure 1.5).

Finally, Jones (1984) has studied the energy budget for the North Sea (an example of the energy budget calculus is shown in Figure 1.6). On the basis of this approach, he concludes that the ascendant demersal-fish abundance in the North Sea is not limited by energy considerations.

It is fair to say that our understanding of the causes of change in the North Sea stocks remains rather opaque. The climatic events are complex and of varying influence and are not likely to be repeated in the future. They are hard to define in an a priori sense and may be scarcely measurable.

Pink Salmon

The complexity apparent in the Pacific sardine–northern anchovy and the North Sea examples may be due to its multiple-species setting. Some single-species settings seem at first glance to be simpler, but more detailed study reveals that single species can also be typified as quite complex. One such case involves the pink salmon; it is specifically included among these examples because of the pink salmon's unique life-history pattern, which raises questions of population control that seem simple and well formulated when compared with those related to other fish species.

Pink salmon are generally anadromous. The adults spawn in rivers and streams of the North Pacific rim (Hanavan and Skud, 1954, however, have provided an example of intertidal spawning). After hatching, the young fish migrate immediately to the sea. Pink salmon remain at sea for exactly two years; those that survive the oceanic sojourn then return to spawn in the stream in which they were born.

The consequence of this two-year life-history pattern is that pink salmon that spawn in even years never mix at spawning with those that spawn in odd years, even though they spawn in the same stream. Thus, remarkably, in any stream the "odd-year line" is reproductively isolated from the "even-year line," as if the two lines were separate species. In addition to the reproductive isolation of odd-year and even-year lines, in a particular stream or river system it is not uncommon for an odd-year line to be consistently much more abundant than an even-year line (see Figures 1.7 and 1.8).

The unique life history of the pink salmon prompts (as did the fisheries for sardine and anchovy off the western coast of North America and herring and cod in the North Sea) a series of questions, the most obvious of which is, why is the odd-year line in any river or river system much more abundant than the even-year line? Is the cause some property intrinsic to each line, or is there some interrrelationship between the two lines that causes one line to maintain a high population level and the other line to be at a relatively low level? Answers to these questions regarding the pink salmon, which does not have a highly sensitive larval

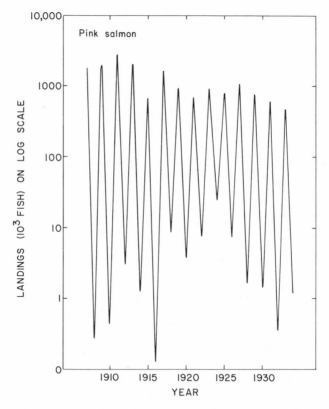

Figure 1.7 Landings of pink salmon from just south of the mouth of the Fraser River. The data points from odd- and even-year-line stocks are connected to emphasize the *apparent* cycles. (Data from Rounsefell and Kelez, 1938: 807; modified after Ricker, 1962: fig. 5.)

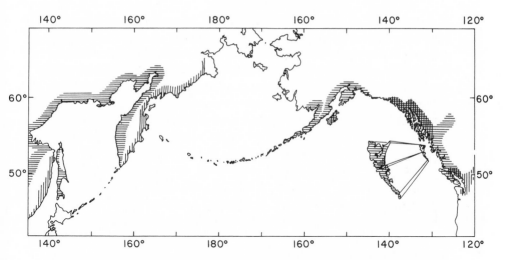

Figure 1.8 Distribution of even-year and odd-year dominance in pink salmon as of about 1925. *Horizontal shading:* even-year dominance; *vertical shading:* odd-year dominance; *cross-hatching:* absence of dominance. No unequivocal information was found for Hokkaido, the southern Kuril Islands, the Anadyr region, Alaska north of Bristol Bay, the Aleutian Islands, or Washington south of Puget Sound. (Inset after Neave, 1952, fig. 1; redrawn from Ricker, 1962: fig. 10.)

stage, might lend considerable insight into the more general aspects of fish-stock variability.

Ricker (1962) has made a detailed analysis of the so-called cyclic-dominance question in pink salmon. His study of the populations of pink salmon over the entire North Pacific rim reflects the complexity of the situation. He stresses three main points. First, dominance exists in varying degrees; in some streams the odd-year line is much more abundant than the even-year line, but in other streams dominance can be less pronounced or virtually absent. Changes in the magnitude of dominance can, however, relate to a diminution in the abundance of the more abundant line but generallly not to an increase in the abundance of the less abundant line. Second, the character of dominance can change. Some regions are even-year dominant while other regions are odd-year dominant; yet entire regions can shift from being odd-year dominant to being even-year dominant. Third, the dominance phenomenon occurred in some regions before the onset of commercial fishing, but in other regions it was observed only after the initiation of fishing.

Ricker suggests eight hypotheses to account for the cyclic-dominance phenomenon: (1) depensatory predation on fry in fresh water, (2) depensatory mortality of fingerlings in salt water, (3) predation between the two lines (cannibalism), (4) fouling of the redds by large egg depositions, (5) depensatory fishing, (6) the influence of density-dependent fishing, (7) competition for food between the lines, and (8) supplementary hypothesis: separation of stocks at sea or along migration routes.

Having sorted out the nature of cyclic dominance and the hypotheses that could account for it, Ricker (1962: 191–192) concludes by noting:

> The pink salmon "cycle" is popularly regarded as an unsolved and unsolvable mystery — something unexpected and unfathomable. To science, of course, there is no problem that is insoluble in principle. But science, as reflected in the customary training of scientists, may be handicapping itself by too strong a predilection for solutions that are at once simple and comprehensive. An hypothesis gains support in proportion to the number of facts it explains, the number of problems it solves. The opposite situation, where one problem may require several hypotheses, is much less acceptable to our conditioning. Yet after many minds have searched for *the* cause of a well-known phenomenon, and when all the hypotheses suggested explain only a part of the facts, it becomes time to consider the merits of multiple causation.
>
> In pink salmon dominance, the basic fact is that the fish mature in their second year of life. Given this, there are several possible natural mechanisms that produce or encourage dominance, as does man's fishing unless it is specifically managed so as to avoid this. I for one am about ready to give up the search for a unique cause of dominance, and concentrate rather on identifying which cause or causes operate on each individual stock. If there is a remaining mystery, it may be this: when dominance can be caused by so many factors, why are there some pink salmon stocks in which it does not occur and has never occurred?

Common Elements

What elements are common to these three examples? In each example there is no shortage of data taken over many years on the status of the stocks and changes in the environment. Many hypotheses have been constructed to explain the decrease or increase in the abundance of particular stocks or groups of stocks, yet the explanations remain tentative and not predictive. In almost all cases the lack of a stable explanation for stock-abundance variability suggests a complex and multiple causation. The concern about prolonged depressed population levels also appears common to the three examples. This phenomenon was evident in the "collapse" of the Pacific sardine and the North Sea herring. The fact that one line of pink salmon tends to remain at a low level of abundance, whereas the other line may exhibit a higher level of abundance in the same stream, has also been of concern, because managers might prefer to increase the abundance of the least-dominant line.

The depressed population levels, particularly the spectacular collapses, were the focus of much attention in the 1960s and early 1970s owing to scientific interest in the causes of the sharp decline and the often severe economic costs suffered by the industries and the fisheries (see, for example, Glantz and Thompson, 1981, regarding the collapse of the Peruvian anchovetta). In fact Murphy (1973) considered this problem in detail and called attention to the "collapse" of a number of clupeoid stocks in addition to the Pacific sardine and the North Sea herring.

Several features of the collapses are worth recording. First, as will be discussed later, collapses are evidenced not simply as a decline in recruitment but rather as radical changes in the production of recruits per unit stock. Second, some collapses *appear* to be irreversible while others appear to be reversible. Third, although much literature has alluded to collapses being a special property of clupeoid stocks, they have occurred in other groups of fish, such as the gadoids (Georges Bank haddock) and the yellowtail flounder in the North Pacific Ocean. Fourth, while it is difficult to demonstrate that collapses are caused by fishing, many collapses do in fact appear to be associated with very heavy fishing. And fifth, while it is difficult to attribute collapses directly to fishing or to determine the causative role of fishing in a collapse, it is also difficult to attribute collapses to environmental causes except in several instances that seem directly related to temperature. (For example, Cushing [1982: 136–145] discusses how the West Greenland cod stock fluctuated with interannual temperature changes. Brongersma-Sanders [1957] presents references on the influence of temperature on mass mortalities. The translocation of stocks on both the east and west coasts of North America due to unusually warm periods is also well known.)

Just as there are unexpected declines — probably not just downward fluctuations but semipermanent reductions in recruitment — there are also unexpected increases or "explosions" in fish populations. These may follow collapses, as in the Atlanto-scandian herring or in the Japanese sardine, but they can also appear de novo. For example, Troadec et al. (1980) and Gulland and Garcia (1984) report a well-known "explosion" of the trigger fish *Balistes* off West Africa. In 1978–79 the catch of trigger fish off Guinea-Bissau was zero; but by 1982, 64 percent of the total landings were trigger fish. Not only did the abundance of trigger fish increase locally, but the increased population spread along the African coastline, apparently paralleling a decline of porgies or sparids, which were heavily fished, implying a "replacement" of porgies by trigger fish. (See Daan, 1980, for a thorough discussion of the "replacement phenomenon.")

Thus, although early attention on population fluctuation concentrated on so-called collapses, there is ample evidence that not all collapses are permanent — at least some are reversible — and in addition populations not only collapse, they also explode. The problem then is not so much one of studying the sudden decrease in population but of studying the change in population.

The difficult problem of trying to explain and predict the changes does not have to be approached completely de novo, however, because, as implied and even demonstrated earlier, numerous hypotheses have been advanced to explain population changes. These might broadly be categorized as follows:

1. *Intraspecific hypotheses* relate to density-dependent changes in mortality, growth and reproductive rates, recruitment-stock relationships, and cannibalism.

2. *Fishing hypotheses* assert that fishing reduces the average population size and that the change in population size causes the annual production of young fish, recruitment, to decrease.

3. *Natural environmental hypotheses* relate to climate- or weather-induced change in the physical structure of the sea or to physical and biological changes per se. The physical changes can relate to changes in the temperature regime or to changes in the biota, such as the changes in calanoid copepods in the North Sea. Change in the natural environment causes the stocks either to decrease or to increase.

4. *Anthropogenic environmental hypotheses* relate to the effects of various pollutants on fish population — for example, the effects of power plants.

5. *Interspecific hypotheses* are the so-called fish-replacement hypotheses (Daan, 1980), which suggest that increase or decrease in one species causes an increase or decrease in another species.

6. *Complexity hypotheses* suggest that the other hypotheses are too simple. Some researchers suggest a specific complexity hypothesis (for example, Skud, 1982).

It is highly probable that any change in the abundance of a fish stock can be accounted for by a mix of the first five hypotheses, which suggests that the best present explanation for fish-stock fluctuation is some form of the complexity hypothesis. The problem, then, is not so much that the first five hypotheses are untenable but rather that the mechanisms they represent operate in a system of very high dimensionality and complexity, which obscures the nature of their interactions. Each of the first five hypotheses seems to describe different parts of the problem with a fidelity that varies as a function of stock magnitude and the "environment" as various conditions change.

Thus the solution of the problem of predicting changes in the abundance of fish stocks will not be advanced by obtaining more of the same data or replicating extant research approaches; rather it will be derived from a reformulation of the problem. This should include a systematic search for those critical parts of the problem that are not well identified and a design of experiments that can expeditiously illuminate these critical areas.

To begin to solve this problem, we must refocus research thinking explicitly to take into account shifting causal mechanisms and specifically to address the following questions:

1. How can we separate the effects of fishing from naturally induced variation?

2. How can we predict the magnitudes of routine fluctuations in recruitment?

3. How can we predict sustained changes or trends in abundance, such as the decline of herring and the increase of cod in the North Sea?

4. How can we appraise the influence of pollution, eutrophication, or habitat modification on the abundance of particular fish stocks?

5. What is the nature of the interaction among species of fish in an ecosystem, with specific reference to predicting a change in abundance of one species as a function of the change in abundance of other species?

2

Examples of Fish-Stock Variability from the Historical Record

The preceding examples of problems in the analysis of fish-stock variability leads us to a more formal examination of its nature and magnitude. At the outset, note that several of the terms describing fluctuations in fish stocks are used in various ways. Following convention, unless otherwise specified the term *abundance* generally refers to the *biomass* of the spawning stock. The term *recruitment* refers to the *number* of young fish that are produced each year; the recruits are identified at a particular age, such as age of maturity or when young fish first become liable to capture by fishing gear. Recruits born in any single year are called an *age class* or a *cohort.* At any given point in time a fish stock consists of several age classes of fish.

The numerical magnitude of each age class is a function of the number of recruits in each age class and the forces of mortality to which they are exposed during their lives. Abundance at any time can be calculated from the product of the number of mature fish and their average weight. Average weight is a function of each individual fish's growth rate. Of the variables of recruitment, growth, and mortality, recruitment is generally assumed to have the greatest influence on the variability of stock abundance.

Analysis of the Historical Record

The value of inferences on future fish-stock variability drawn from historical records depends on the extent to which the historical data actually represent either past or future sequences of annual recruitment data. Major issues regarding the representativeness of historical data include: (1) the relation of interannual variability in recruitment to interannual variability in abundance; (2) the extent to which the available sample of recruitment data can be considered "typical" of recruitment variability in marine fishes; and (3) the extent to which the relatively short sequences of available recruitment data can be considered to represent longer time series.

Relation of interannual variability in recruitment to interannual variability in abundance. It is important to recognize that variations in recruitment affect characteristically high-mortality-rate stocks (such as skipjack tuna) differently from characteristically low-mortality-rate stocks (such as cod). The former have relatively few extant age groups, whereas the latter have relatively many; hence recruitment variability would have a greater effect on the overall abundance of typically high-mortality stocks than on low-mortality stocks, all other things being equal. In other words, fluctuations in abundance of low-mortality stocks should be smaller than those of high-mortality stocks.

Representativeness of the sample of available data. Tens of thousands of species of marine fish exist, but the sample of recruitment time series is probably no larger than a few hundred species. The available data may represent the kinds of variability to be expected in the highly fecund marine fish that are taken in commercial fisheries. It is not certain, however, that the variability observed in these samples is typical of other marine fishes. Nevertheless, the essence of long-term recruitment variability can be captured from a consideration of the few known long-term records (at least one century long) and some of the sequences from well-studied areas.

Degree to which the relatively short time series of available recruitment data represents long-term variability. Very few sequences of recruitment data extend back more than fifty years from the present. To put this in perspective, consider the actual sequence of recruitment and compare it with a typical time series of available data. The actual sequence can be represented by

$$\mathbf{R} = R_0, \ldots, R_N, \tag{2.1}$$

where R_i is the annual numerical magnitude of recruitment in the ith year (for $i = 0, \ldots, N$) before the present year, and N is a very large number.

The actual time-series data may be represented by

$$\mathbf{r} = r_0, \ldots, r_n, \tag{2.2}$$

where r_i (for $i = 0, \ldots, n$) is an estimate of corresponding R_i, and n is the number of years for which recruitment data are actually available.

The point of writing (2.1) and (2.2) is to emphasize the considerable difference between them. Even given that the estimates in (2.2) are perfect (see, for example, Walters and Ludwig, 1981, and Ludwig and Walters, 1982, on measurement errors), n is generally considerably smaller than any reasonably small value of N. Further, the sequence $\mathbf{r}$ is not necessarily representative of the time interval represented by $\mathbf{R}$, because $\mathbf{r}$ is based only on the n most recent years of N.

Much of the motivation for studying recruitment involves the desire to predict future properties of $\mathbf{R}$ (that is, $\mathbf{R'} = R_{-1}, R_{-2}, \ldots$) based on $\mathbf{r}$ and the particular biotic or abiotic conditions that might be associated with each particular r_i. Factors important to predicting $\mathbf{R'}$ from the observations of $\mathbf{r}$ include (1) the degree to which $\mathbf{r}$ represents $\mathbf{R}$, (2) the environmental or other conditions associated with each r_i, and (3) the degree to which these conditions might be associated with future values of r_i.

Having cited the nature of the relation between the long-term recruitment sequence and the recent historical record, we can now examine some of the logical problems that arise in drawing inferences from the

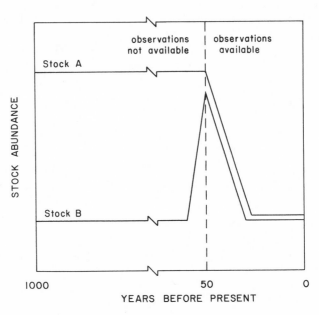

Figure 2.1 Trajectories of abundance for two hypothetical stocks A and B. Observations are available only within the past fifty years. Observations suggest that both stocks collapsed, but in actuality only stock A collapsed, whereas stock B returned to its normal level of abundance.

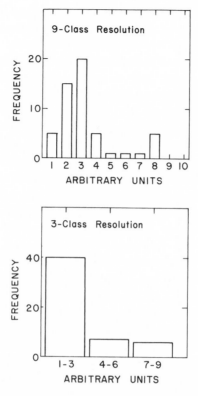

Figure 2.2 Two histograms based on identical arbitrary data showing the effect of resolution on the shape of the distribution. The differences in resolution could be due to the classification scheme or to measurement errors.

available data and the historical record. This question is of fundamental importance, because a particular prolonged increase or decrease in annual recruitment can represent two very different circumstances that cannot be distinguished simply by inspecting the time-series data. For example, a population collapse may be a serious permanent or semipermanent reduction in population level. Yet a collapse may simply reflect the fact that at the beginning of the observational period the stock was at an abnormally high level of abundance, and so the reduction in abundance is not a collapse but simply a return of the population to its normal level of abundance.

These two situations are depicted in Figure 2.1. It is easy to see that *all* causal mechanisms asserted to be associated with observed fluctuations in animal populations are subject to question unless there is some way of distinguishing between a "normal" population level and an "abnormal" population level. That is, the description of the effect cannot be made from the time series alone; without some underlying theoretical basis, it can be as complicated as assigning the cause.

Another problem that arises in attempts to synthesize or analyze recruitment data concerns the frequency distribution of recruitment values. The form of this empirical distribution is a function of the way in which individual observations are grouped in size classes. The grouping can reflect a natural grouping of the data, but it can also reflect measurement error. In either case, frequency tabulations could result in sample histograms whose shapes can be different from those of "true" histograms (Figure 2.2).

Long-Term Record

The relatively few long-term sequences of stock-abundance data are based on our knowledge of the paleontological record, on more or less anecdotal or qualitative medium-term information, and on analyses of sediment containing centuries-old depositions of fish scales. More precisely, the long-term record tends to relate to variability in stock abundance more than to variability in recruitment; nevertheless, it suggests the nature of recruitment variability.

The first living organisms appeared 3.5 billion years ago. The fishes appeared much more recently in the Mesozoic era, some 135 million years ago. Our understanding of the fossil record and of evolutionary processes suggests that many species have become extinct, while new species have differentiated from ancestral species. Thus, over the course of its existence, each extinct species must have had at least an initial period of increase, a period of relative stability, and a period of decline.

With respect to medium-term information, a score of anecdotal records and reports of fluctuations in fish abundance have appeared over the past two or three centuries. Many of these have been reviewed by Cushing (1982). As an example of one picturesque report, Braudel (1981: 215) writes:

> A Dutchman, William Beukelszoon, is said to have discovered in about 1350 the rapid method of gutting herrings and salting them on the boat where the fishermen could barrel them immediately. But the herring disappeared from the Baltic between the fourteenth and fifteenth century. After that, boats from Holland and Zealand fished on the barely covered sands of the Dogger

Bank and in the open sea off the English and Scottish coasts, as far as the Orkneys. Other fleets gathered at these rich grounds. In the sixteenth century at the height of the conflicts between the Valois and Hapsburgs, herring truces were duly concluded to ensure Europe's continued supplies.

Commonly used references include Uda's (1961) presentation of information on periods of scarcity and abundance of several species of fish important in Japan, dating back to the seventeenth century, and Devold's (1963) information on the catches of herring from the Bohuslan, northern and western Norway, dating back to 1760.

In addition to the anecdotal records, a few examples of long-term records exist, based on fish scales obtained from varved cores taken from anoxic sediments. A time series developed by Soutar and Isaacs (1969, 1974) shows a periodic disappearance of sardine, a steady decline in anchovy, and periodic changes in the abundance of hake. While hake appear to be omnipresent, there seem to be periods when sardine are at a zero level of abundance (see also McCall, 1983; Lasker and McCall, 1983). Figure 2.3 is developed from time-series material presented by Soutar and Isaacs (1969) for sardine, anchovy, and hake in the Los Angeles Basin. The figure shows the extent of correlation among the abundances of the three species. Hake and sardine do not occur together more often than hake and anchovy; if anything, there seems to be a slight positive correlation. It is not clear to what extent these relationships can be attributed to actual changes in abundance or to movements of the center of distribution of fish owing perhaps to warming or cooling, related to El Niño, southern oscillationlike events. The Soutar and Isaacs papers are frequently cited as representing the dynamics of the sardine, hake, and anchovy complex off southern California. But it is not often stressed that the data are based on spatial samples that may not be representative of the actual abundance of the three species; very different conclusions could be reached if significant variations in scale deposition were found in different locations. In addition, any temporal variation in the spatial position of the mix of species known to occur in modern times as a result of the El Niño phenomena would cause the apparent local abundance of the three species to fluctuate when the actual abundance was constant. Additional evidence on the long-term-variability interval from depositions of fish scales is presented by DeVries and Pearcy (1982).

These rather sketchy long-term records show, then, that fish stocks have been fluctuating in abundance for centuries: stocks have increased and decreased and exhibited considerable variation in abundance in the absence of fishing and other anthropogenic activities.

Figure 2.3 Correlations in abundance (*a*) between Northern anchovy and Pacific hake, (*b*) between Pacific hake and Pacific sardine, and (*c*) between Northern anchovy and Pacific sardine. (Data from Soutar and Isaacs, 1969: figs. 7, 8, 9, based on scale depositions for the past two millennia.)

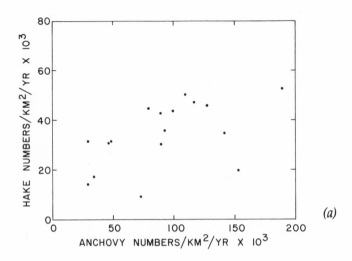

(*a*)

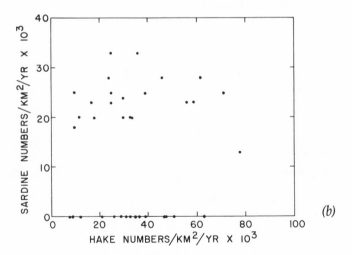

(*b*)

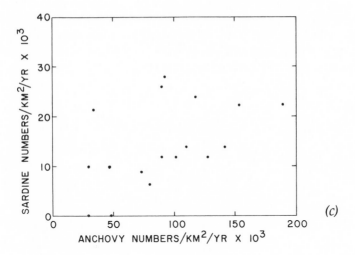

(c)

Short-Term Record

As might be expected, relatively more sequences represent the short-term record than the longer-term record. Garrod (1982) has, for example, presented short-term recruitment time series from a number of well-studied stocks. The data presented by Garrod (Figure 2.4) represent fairly well the range of variability that might be expected in fish stocks.

Recruitment trends can be discerned in Figure 2.4. The data set does not appear to reflect obviously increasing trends; some populations appear to reflect constant but fluctuating recruitment, while for other stocks there is a steady decline in recruitment. Declining trends are evident for the California sardine, the St. Lawrence cod, the Greenland cod, the North Sea herring, and the Norwegian spring-spawning herring. The North Sea plaice, the arcto-Norwegian cod, the Northeast arctic haddock, the North Sea haddock, and the Georges Bank haddock exhibit no obvious temporal trends in annual recruitment.

In addition to recruitment trends, we can discern variability patterns. Figure 2.5 shows the frequency distribution of year-class strength for each species. We can see that the distributions tend to be asymmetric, skewed to the right — that is, most recruitment values are relatively low. These relatively low values are interspersed by the occurrence of what might be called very large year classes (VLYC). The distribution of a relatively few VLYC among smaller year classes has been interpreted as reflecting the fact that recruitment time series are drawn from a log normal probability distribution (Hennemuth, Palmer, and Brown, 1982).

Figure 2.4 (a–k) Time series of recruitment of various species of fish. (Data from Garrod, 1982: app. 1, table 1, plotted as percent deviations from mean recruitment.)

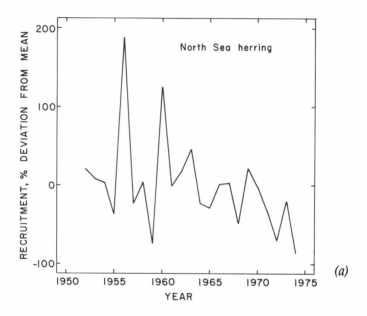

(a)

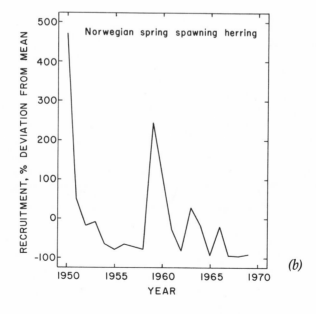

(b)

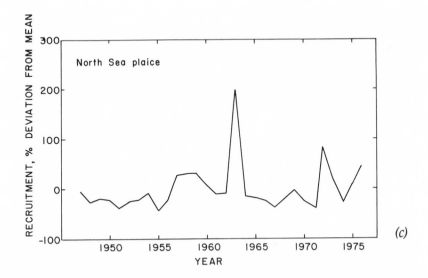

(c)

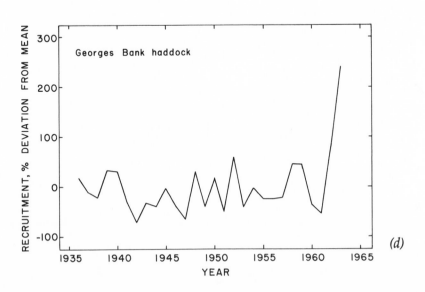

(d)

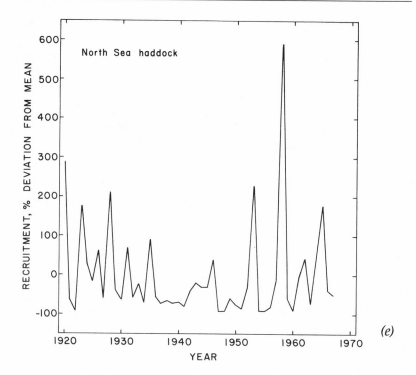

(e)

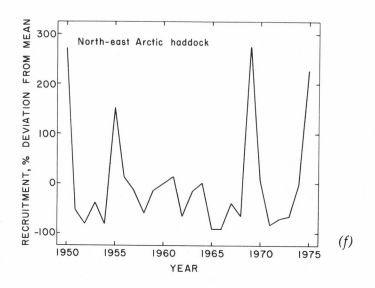

(f)

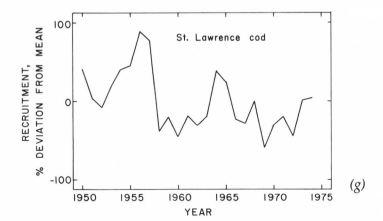

(g)

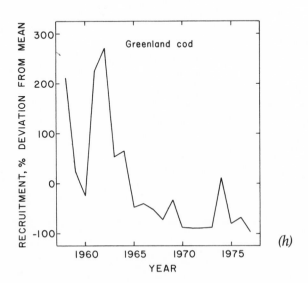

(h)

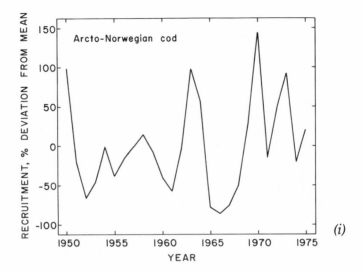

(i)

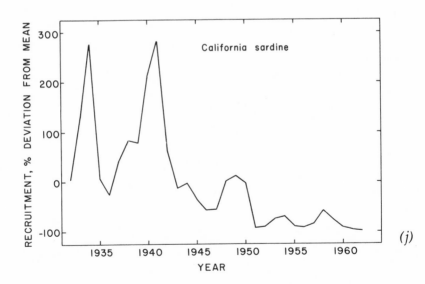

(j)

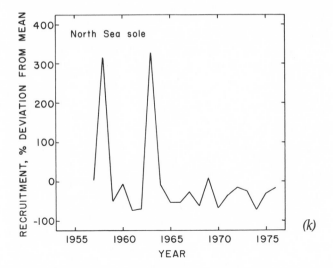

(k)

But the declining trend, often present in recruitment statistics, if not removed, would contribute to increasing the proportion of smaller year classes or conversely decreasing the proportion of larger year classes, which may contribute to the impression of skewness. Further, the frequency distributions in Figure 2.5 do not appear to be log normal; they appear to be contagious (see Feller, 1957, for a contrast between real and spurious contagion). The contagion seems to result in part from the VLYC but also from the fact that after removing the VLYC from consideration, recruitment values seem to fluctuate between maximum and minimum values with relatively few intermediate values.

The range of maximum to minimum recruitment values is an index of the magnitude of recruitment variability that might be expected from different stocks. These values are given in Table 2.1, along with information on VLYC. Some caution needs to be applied to interpreting maximum-minimum ratios, because most statistics of "extremes" (that is maxima, minima, and ranges) are, as one might expect, dependent on sample size (Gumbel, 1959). For instance, larger extremes can sometimes be expected with larger sample sizes. Further, the maximum-minimum ratio can escalate quickly when, over time, there has been at least one very small minimum value. Hennemuth and colleagues (1982) give the maximum-minimum ratio of recruitment size for three different time periods for Georges Bank haddock. For 1931–1965, when population was at a relatively high level, the ratio was 19:1; for 1966–1973, when the population was at a low level, the ratio was 92:1; but when both

Table 2.1 The ratio of maximum recruitment to minimum recruitment and the number of years per very large year class (VLYC) for various stocks.

Stock	Number of years	Max. recruit $\times 10^6$	Min. recruit $\times 10^6$	Max./ min.	Number of VLYC	Years per VLYC
California sardine[a]	31	21,545	110	196	2	15
Norwegian spring-spawning herring[a]	20	78,300	600	130	2	10
North Sea herring[a]	23	21,370	1,040	21 (10)	2	12
St. Lawrence mackerel[a]	10	7,514	1,710	4	1	10
Arcto-Norwegian cod[a]	26	2,920	170	17	4	6
Greenland cod[a]	20	718	7	100	2	10
St. Lawrence cod[a]	25	214	62	3	4	6
North Sea haddock[a]	48	6,297	55	114	5	10
Georges Bank haddock[a]	27 (43)	283	23	12 (2,700)	1	27
North Sea plaice[a]	30	1,310	249	5	1	30
North Sea sole[a]	22	598	38	16	2	11
Northeast Arctic haddock[a]	26	1,540	30	50 (9)	4	6
Georges Bank cod[b]	14	—	—	2.5	1	14
North Sea cod[b]	15	—	—	5.5	3	5
Georges Bank herring[b]	15	—	—	5	2	7
North Sea mackerel[b]	10	—	—	41	1	10
Northwest Arctic mackerel[b]	12	—	—	18	1	12
North Sea saithe[b]	18	—	—	12	2	9
North Sea whiting[b]	16	—	—	7	4	4
South American pilchard[b]	16	—	—	10	2	8
South American anchovy[b]	13	—	—	3	3	4
Round herring[b]	13	—	—	10	1	13
Peruvian anchovy[b]	16	—	—	11	2	8
Silver hake[b]	19	—	—	10	1	19

a. From Garrod, 1982.
b. From Hennemuth, Palmer, and Brown, 1982.

periods are combined, the ratio is 2700:1. Consider also the Peruvian anchovetta, which had a maximum-minimum ratio of 11:1 during the 1961–1976 period. (If the most recent years of 1983–1984 were utilized in this calculation, the ratio would increase by orders of magnitude.) Keeping these cautions in mind, and taking into account the irregular

Figure 2.5(a–j) Frequency distributions of recruitment of various species. (Data from Garrod, 1982: app. 1, table 1.)

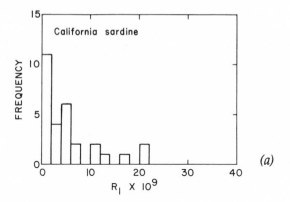

(a)

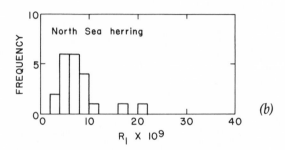

(b)

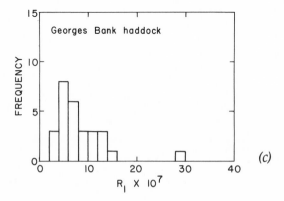

(c)

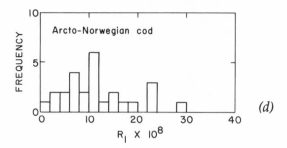

(d)

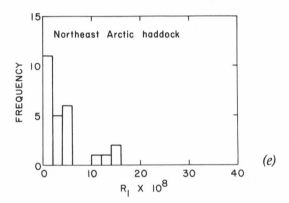

(e)

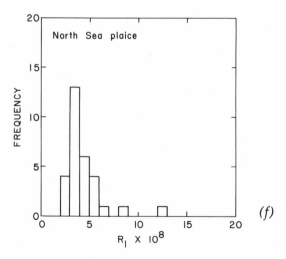

(f)

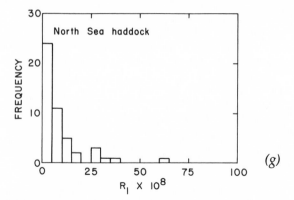

(g)

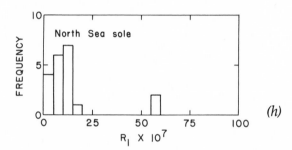

(h)

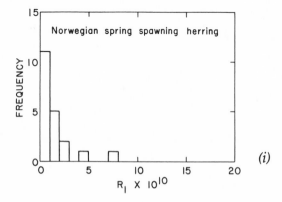

(i)

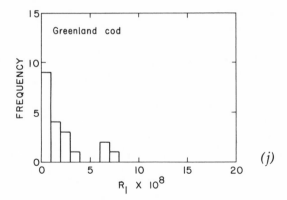

(j)

form of the frequency distribution, we find that the maximum-minimum ratio is nevertheless a handy index of population variability.

The number of VLYC are also presented in Table 2.1. What is or is not a VLYC is somewhat subjective. The number of VLYC was judged by visual examinations of Figure 2.4 and the material presented in Hennemuth et al., which led to the conclusion that, on average, VLYC are not common. They occur in the sample in Table 2.1 an average of once every eleven years.

With regard to the variation, we can observe the following:

1. In any stock it should not be surprising to find maximum-minimum values of 10 or 20. A few stocks have been very highly variable, exhibiting ratios greater than 50. (It is interesting to observe that during the much-heralded collapse of the Peruvian anchovetta the ratio was only 11 : 1.)

2. The mean interval between VLYC is about eleven years. (This is, of course, not to say that we should expect a VLYC during an eleven-year period, but rather that VLYC do not occur very often; we might expect to see one once every eleven years or so.)

3. Of the groups of fish listed, no particular group (such as gadoids, scombrids, clupeoids) seems to be any more or less variable than the others.

4. The results indicated above are based on a somewhat subjective analysis and are particularly sensitive to the number of years for which data are available.

To summarize, it is possible to classify annual recruitment time sequences into five categories:

1. Stocks in which the magnitude of recruitment fluctuations seems relatively stable and in which the recruitment in each year appears to be statistically independent of recruitment in any other year. There may be no trend in average recruitment or there may be an increasing or decreasing trend.

2. Stocks in which the magnitude of recruitment fluctuations seems relatively stable and in which the recruitment in each year seems to be serially correlated with recruitment patterns in other years. Recruitment in these stocks may be strongly cyclical or weakly cyclical. Again, there may or may not be a trend in average recruitment.

3. Stocks whose recent recruitment patterns are dominated by one or two very large year classes that are evidenced in the sequence of recruitment data.

4. Stocks whose recruitment pattern is reflected not in a decreasing trend but in a sharp and sustained downward fluctuation.

5. Stocks whose recruitment pattern is reflected not in an increasing trend but in a sharp and sustained increase in abundance.

Of course, in biology there is no perfect classification scheme. The recruitment pattern of some stocks will fit one of the five classes perfectly, while that of others will be difficult to categorize, and some stocks may be placed in different categories at different points in their history. The recruitment from various stocks can then be examined, and we can attempt to classify each stock into its appropriate pattern.

A consideration of the recruitment-variability record enables us to ask more precise questions as to why stocks exhibit a particular type of variability.

3

Introduction to Fishery-Related Population-Dynamics Theory

Trends, collapses, explosions, and various periodicities are evident in time series of recruitment data. Despite the variability, however, fish stocks appear to be remarkably stable in the sense that, to the best of our knowledge and within our relatively short time frame of interest, they are neither becoming extinct nor increasing without check.

A frequent reaction to observations of the magnitude or behavior of fish-stock variability is to ask why fish stocks vary (see, for example, Sissenwine, 1984). A less common reaction is to ask why fish stocks evidence their remarkable stability (see, for example, Ursin, 1982; Gulland, 1982).

Pursuit of answers has generally been channeled through an examination of recruitment time series. Unfortunately, the consequence of such an approach is that conclusions derived from empirical examination of recruitment time series must be circumscribed by the information contained in the time-series data. Because most recruitment time series are relatively short, they are unlikely to contain events that occur only occasionally, such as collapses, leading the analyst to believe that collapses are either more or less likely than they actually are — an example of the difficulty inherent in attempting to predict the occurrence of unusual events from a short time series. If this information cannot be derived from available time series or observations of past events, then it must be derived from a better theoretical understanding of the recruitment process.

It is convenient and natural to use extant population-dynamics theory as a basis for developing a better theoretical understanding of the process. The theory has as its essential aim the prediction of temporal and sometimes spatial variations of fish-population numerical abundance, taking into account reproduction, growth, and mortality as well as the environmental variables that affect these vital rates. This theory, its strong points and its weak points, can be best appreciated in the context of unique properties of fish-population dynamics.

Unique Properties of Fish Populations

In developing the context for considering fish-population dynamics, it is important to identify properties of fish populations that tend to set fish apart from other vertebrates. These include (1) production of a superabundance of eggs, (2) external fertilization, (3) a considerable difference between the scale of events associated with adult fish and that associated with very young fish, (4) an "environment" that is abstract and difficult to define, and (5) an "external" force of mortality: fishing (other vertebrates are, of course, subject to various forms of human predation).

Production of a Superabundance of Eggs

Many marine fish produce a "superabundance" of eggs (the term is not intended to imply that there are more eggs than necessary but to signify that many marine fish spawn a very large number of eggs). In contrast, most vertebrates produce relatively few eggs or young per female. For example, mammals produce several young per litter; birds produce several eggs per clutch; reptiles tend to produce somewhat more young per season; and the order of magnitude of the production of eggs or young in amphibians is only in the hundreds or thousands. It would not be uncommon, in many species of marine fish, for an individual female to spawn several hundred thousand eggs in a single spawning season. If only a few are required to survive to maturity to maintain a stable population, why do marine fish spawn such very large numbers of eggs?

It must be recognized that there are varying degrees of egg superabundance. The marine fish taken in the major commercial fisheries produce the largest numbers of eggs. Among all species of fish, it appears that declines in fecundity are coupled with increased parental care (for a discussion of the evolution of parental care in fish, see Gross and Sargent, 1985). Further, while some fish seem quite abundant as eggs and adults, others are extremely abundant in the plankton and hardly noticed as adults — perhaps because they are not caught commercially or because their life history is different in the sense that major mortalities occur after these species are evident in the plankton.

External Fertilization

Fish are the only higher vertebrates with generally external fertilization. In other higher vertebrates fertilization requires male-female contact and occurs under relatively stable conditions; but in the case of fish, the gametes are ejected into the ocean, subjecting the fertilization process to the vagaries of ambient chemistry, temperature, light, and motion. It is not clear whether the large numbers of dead eggs that are observed from time to time result from cytological incompetence, a failure in the fertilization process, or postfertilization death.

Heterogeneity of Scale

Fish populations are characterized by considerable differences in the scale of events that occur when the fish are very young as opposed to when they are juvenile or mature. These scales relate to (a) mortality rates, (b) growth rates, and (c) ambit per unit time.

Adult fish die at an instantaneous rate of perhaps 0.5 per year; in contrast, larval fish die at an instantaneous rate of perhaps 36 per year (about 10 percent per day). The magnitude of egg and larval mortality might be visualized by noting that we would observe the death of one larva each minute in a relatively small batch of fish larvae — say, ten thousand — dying at a rate of 10 percent per day. Because the time span of the life-history stage is inversely proportional to the mortality rate, the relatively short time scale of events relevant to the larval stage can be easily seen. (Figure 3.1 shows the importance of the days-weeks scale in various aspects of the life history of larval fish; see also Kawai and Isibasi, 1983.) With respect to size-specific growth, adult fish grow very little, and young fish grow at a much faster rate, possibly quintupling their length in a month or so (even greater changes in weight might be expected). With respect to ambit, individual adult fish can swim at relatively high velocities, while larval and postlarval fish swim at velocities that enable them to search only a liter or less per hour.

The greatly varying scale of events associated with larval fish raises the question of appropriate temporal and spatial sampling frequencies, particularly for very young fish. Because of the microscale and fine-scale nature of their environment, it is important to sample not only at temporal and spatial frequencies that capture the essence of environmental change in the fish, but also at times of the year when these critical events are occurring.

Further, it is clear that the microscale and fine-scale events are nested in a matrix of phenomena existing on larger scales, so we must try to understand the extent to which linkage occurs between the microscale and fine-scale events and the macroscale events. This linkage may be seen by modifying Steele's 1978 presentation of the linkage among plants (that is, phytoplankton), herbivorous zooplankton, and pelagic (adult) fish (see also Stommel 1963, 1965; Haury, McGowan, and Wiebe, 1978). Steele's diagram can be modified to show the time-and-space

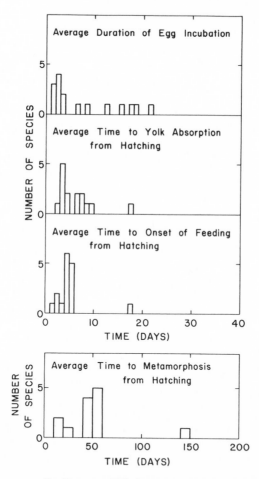

Figure 3.1 Frequency distribution of life-history events in a variety of species of fish. These data show the relatively short time scales and sizes associated with young fish. (Data from Thielacker and Dorsey, 1980; for additional data see Kawai and Isibasi, 1983.)

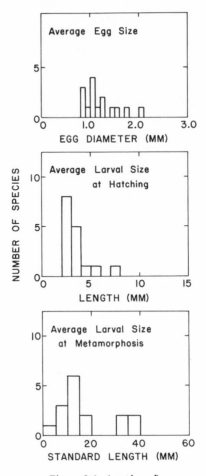

Figure 3.1 (continued)

scales associated with both fish eggs and very young fish, which tend to be approximately in the zooplankton time-space regime and the fish time-space regime (Figure 3.2).

Identification of the nature of the connection between the microscale and fine-scale events is one of the most pressing problems in marine biology. Rothschild and Rooth (1982: 10–11) observed:

Events that affect the survival of larvae operate in the micro- and fine-scale. Yet, for any single recruitment year, there must be trillions of micro- and fine-scale events entrained in relatively large time-space scales (weeks and tens-of-miles). Thus, mechanisms must exist which represent an integration or averaging of the micro- and fine-scale events. The nature of these mechanisms may be one of the key questions which, when answered, will lend considerable understanding of the dynamics of the micro- and fine-scale ecosystem.

Put another way, observations taken at larger scales result from the averages or integrations at smaller scales. In order to interpret the larger-scale observations, it is necessary to understand relations among scales and integrating or averaging weights.

Identification of the Environment

If the dynamics of fish are to be understood, it is not sufficient to consider only the abundance of a particular stock of fish per se; we must also consider the environment in which it lives, because the environment has a significant effect on the dynamics. But what is this "environment"? The marine environment is fundamentally more abstract and more difficult to

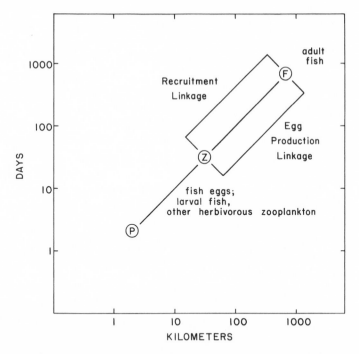

Figure 3.2 Superimposition of fish-population dynamics on Steele's diagram (1978: fig. 3). Typical time and space scales associated with plants (*P*), herbivorous zooplankton (*Z*), and pelagic fish (*F*).

describe than the terrestrial environment. While the terrestrial environment can easily be characterized by a structure that is often obvious, the marine environment is essentially featureless, at least in an anthropomorphic sense. There are, of course, gradients in chemicals and physical composition, but these are often changing and ephemeral, whereas those of the terrestrial environment are tangible.

It is easy to make a variety of at-sea measurements of physics and biota. The question often remains, however, whether the variables that are measured are either important or measured on the right scale. Indeed, there is no guarantee that variables accounting for a major portion of recruitment variation in one year on particular scales will account for a major portion of variability in subsequent years and on the same scales, adding further to the notion of the marine environment's large and complex dimensionality and abstractness.

Thus the critical question regarding the environment is how to make it less abstract and, in so doing, to design the set of measurements relevant to identifying and analyzing the driving forces of population variability. One approach to this problem has recently been presented by Andrewartha and Birch (1984).

Fishing Mortality

The level of fishing mortality varies greatly among fish populations. While there is evidence that fishing can affect the density of the population and its biomass, there is no evidence, other than circumstantial, that fishing has actually caused any long-term documented population trends. The critical question then relates to the extent to which fishing actually affects stocks.

Any explanation of fish-stock variability needs to be set in the contextual framework that makes marine-fish populations unique. The conceptual framework should include the superabundance of eggs and the fertilization process; the kaleidoscope of time-space scales in which the fish live; the difficult-to-define highly dimensional environment; and the role of fishing mortality.

Survey of Population-Dynamics Theory

The essential aim of population-dynamics theory is to predict variations in numerical abundance of fish populations. This section focuses on those aspects of the theory that account for the transformation of eggs into larvae, juveniles, adults, and eggs, completing the dynamic process. (A fairly complete articulation of the theory and methodology for estimating component parameters may be found in Beverton and Holt, 1957; Ricker, 1975; and Cushing, 1981).

The dynamics is usually thought of in terms of two branches of theory. One branch involves the elaboration of biomass in a cohort as a function of time, mortality, and growth, and the transfer of cohort biomass to the next cohort as a function of initial cohort biomass. The elaboration of cohort biomass is addressed in yield-per-recruit theory, and the transfer of biomass from extant cohorts to subsequent cohorts is addressed by stock-and-recruitment theory. The second branch integrates mortality, growth, and reproduction rates, deriving a relation between the rate of change of biomass and biomass. This relation is called production-model theory. The following discussion surveys the elaboration of biomass and stock and recruitment and then production-model theory.

Elaboration of Biomass and Stock and Recruitment

Elaboration of biomass and stock and recruitment are best conceived in terms of specific methodological stages. The methodology includes (1) defining the initial number of recruits, (2) establishing the number of fish that are alive at any time or age after the time or age at recruitment, (3) establishing the weight of fish at any age, (4) establishing the biomass of the group of recruits at any time or age, (5) establishing the number of eggs produced, and (6) establishing the transformation of the biomass of extant cohorts into future cohorts by using stock-and-recruitment theory.

Defining the initial number of recruits. The group of fish born in a single year is called a *cohort* (Figure 3.3). Establishing the notion of a cohort simplifies analysis because it enables us to study the evolution of growth and mortality of a single group of fish, all of the same age, rather than doing the more difficult task of studying all age groups simultaneously. In fisheries terminology a cohort is usually referred to as a year class. For example, in 1982 a fish population might consist of three cohorts: the 1982 year class (young-of-the-year), the 1981 year class (one-year-olds), and the 1980 year class (two-year-olds).

While most theory identifies a cohort as a group of fish born in a single year on a particular date or during some short time interval, deviations from this definition are not uncommon. Instead of spawning on a more or less fixed date, some species of fish spawn continually over much of the year, while others spawn at several different times during the year.

A beginning step in tracing the abundance and mortality of a cohort of

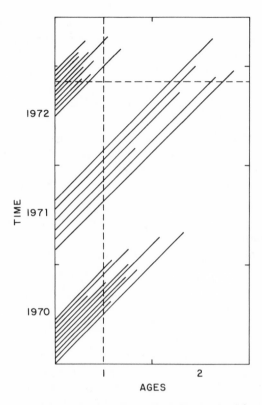

Figure 3.3 The evolution of three cohorts. Each line is the life span of a single fish. Lines represent the longevity of fish born in 1970, 1971, or 1972. Note that the mortality rate in 1971 is less than the mortality rate in either 1970 or 1972. The mean age, of course, decreases as mortality rate increases. The age frequency may be obtained at any fixed time. For example, the dashed horizontal line shows the age frequency of the individuals in the population during a day in 1972. The dashed vertical line shows the number of one-year-old fish alive in each of the three years. We can also transform time into length or weight. (Figure modified after Skellam, 1971.)

fish through time is to establish the initial number of fish in the cohort. (There are many approaches for determining the initial number of fish. One involves direct estimation by using sampling programs to estimate the number of juvenile fish. Another uses catch statistics to obtain an index of abundance, usually in terms of catch of young fish per unit of fishing effort. A third approach uses the catch equation and the number of fish caught at each age to estimate the numbers of young fish caught each year; this procedure is frequently called cohort analysis.) The very high fecundity and the extremely high and variable mortality rates during the first few months of life make it advisable to calculate the initial number of fish in a cohort at a point in time after the initial high-mortality rate subsides. This point in time is often called the age at recruitment; the term *recruitment* refers to the number of fish in the cohort that are alive at the age or time of recruitment. Thus recruitment can be either (1) the number of fish alive in a population at any arbitrarily defined point in time after the subsidence of the initial high mortality, or (2) the number of fish alive in a population at a specific arbitrarily defined point in time, such as when the fish first become available to capture or are first taken in the fishery.

Establishing the number of fish alive after the age of recruitment. Having established an initial number of fish in the cohort, N_0, we can set the stage for computing the number of fish alive at any time after recruitment. Owing to mortality, the number of recruits or the number of fish in the cohort decreases as time increases. The elementary theory that represents the decrease is well known (see, for example, Ricker, 1975) and is based on the assumption that the rate of decrease is proportional to the number of individuals present in the population, namely,

$$\frac{\Delta N}{\Delta t} \propto N, \tag{3.1}$$

where N is the number of individuals in the population and t is time.

If $\Delta N / \Delta t$ is considered as a derivative and $-Z$ is taken as the constant of proportionality, (3.1) has the well-known solution

$$N_t = N_0 e^{-Z(t-t_0)}, \tag{3.2}$$

where N_t is the number of fish in the population at time t, and N_0 is the number of recruits at the recruitment time t_0. For expositions on mortality rates, it is frequently convenient to take logarithms of (3.2):

$$\log N_t = \log N_0 - Z(t - t_0), \tag{3.3}$$

because with constant Z, the decline in the log N_t is linear with time. In Figure 3.4 the interval $t_0 < t < t_r$ is subject to variable and relatively intense mortality. In practice there may be little relation between N_0 and N_{ri}. Clearly the slope of the line for $t > t_r$ is equal to Z; hence cohort 2 has the highest mortality rate, and cohort 3 the lowest. Plausible complications arise when t_r is different for each year or when Z, which is typically thought of as $Z = Z(t)$ becomes $Z = f(t)$. Equations (3.1)–(3.3) are used in many fields other than fisheries. (See Ricker, 1958, for an exposition and Rothschild, 1967, for a stochastic derivation of mortality rates.)

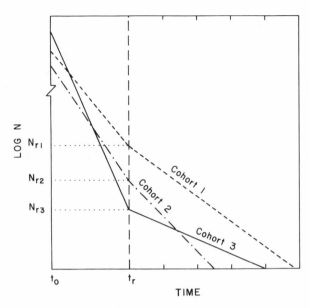

Figure 3.4 Relation between log N and time (or age) for three cohorts. The symbol t_r is the time of recruitment, and N_{ri} is the number of recruits in the ith cohort.

The total mortality coefficient Z is usually taken as the sum of the natural mortality coefficient, usually denoted by M, and the fishing mortality coefficient, usually denoted by F. Hence we can replace Z with $M + F$ whenever necessary. In the simplest instance F is proportional to nominal fishing effort where the constant of proportionality q is called the catchability coefficient (Rothschild, 1977, gives a detailed discussion of the interrelationship between fishing effort and fishing mortality). Normally most discussions of fish-population dynamics place early emphasis on the effect of fishing and the fishing mortality coefficient, but in our discussion emphasis on fishing mortality is suppressed in order to emphasize the importance of natural changes.

Establishing the weight of fish. Establishing the change in average weight of a fish in a population as a function of its age or time might be based on a variety of theories that attempt to account for growth, usually derived from energetic considerations. In reviewing these theories, Ricker (1979) finds that although age, nutrition, and temperature can be linked empirically with growth, a mathematical theory expressing these relationships is elusive. Of the relationships used in the past, the von Bertalanffy growth function has been particularly popular (see Beverton and Holt, 1957). Ricker notes that it is more appropriate to call the relationship the Pütter growth curve. The von Bertalanffy or Pütter growth function can be derived by assuming that the rate of growth is proportional to the difference between the length of the fish l_t and a maximum asymptotic size L_∞. Taking K as the constant of proportionally yields the derivative of length with respect to time:

$$\frac{dl}{dt} = K(L_\infty - l_t),$$ (3.4)

which has the well-known solution

$$l_t = L_\infty(1 - e^{-Kt}).$$ (3.5)

This equation, Pütter's equation 1, which expresses the growth of fish in length as a function of time, can be transformed into an equation for weight by cubing both sides and redefining K to obtain Pütter's equation 2,

$$W_t = W_\infty(1 - e^{-Kt})^3,$$ (3.6)

where W_∞ is the average maximum weight of the fish.

Establishing the biomass. Having established the number of individuals in the cohort at any age or time and the average weight of an individual as a function of its age or time, we can now obtain the biomass of the cohort at any age or time by multiplying (3.2) and (3.6):

$$N(t)W(t) = B(t) = N_0 e^{-Zt} W_\infty(1 - e^{-Zt})^3.$$ (3.7)

The function $B(t)$ first increases, then reaches a maximum at a point in time called the critical age, and finally declines. The shape of $B(t)$ and the position of its maximum is determined by Z, W_∞, and K. Data on the values of M, K, and L_∞ (or W_∞) have been tabulated for many species of fish (Beverton and Holt, 1959; Beverton, 1963; Pauly, 1980). It appears that natural-mortality rates for each species tend to be roughly specific to the family to which the fish belongs (Figure 3.5). For example, some of the flatfishes (Pleuronectidae) have a relatively low natural-mortality rate, whereas some of the tunas and mackerels (Scombridae) have an unusually high natural-mortality rate. These rates imply that the average age of fish in an unfished population belonging to the family Pleuronectidae might be about ten years, while the average age of fish in a species belonging to the family Scombridae might be less than a year.

Cushing (1981) has observed, along with Beverton and Holt (1957) and Beverton (1963), that there appears to be a relation between M and K

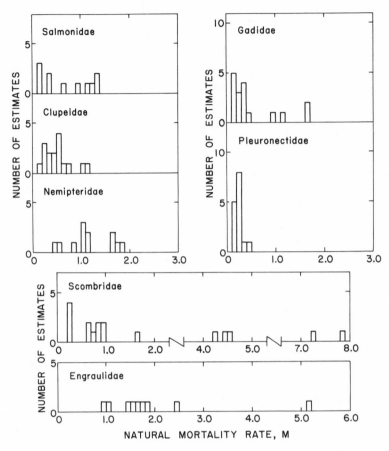

Figure 3.5 Frequency distribution of natural mortality rates. (Data from Pauly, 1980: table 2.)

that is also family specific; for example, it appears that for herrings M is slightly greater than K, but for gadoid fish M is roughly double K.

To demonstrate the effect of different values of M, K, and W_∞, the biomass curves for cod, herring, and skipjack tuna have been plotted in Figure 3.6, using in each case an initial recruitment of 10^3 fish. We can see a substantial effect. For example, the elaboration of biomass of cod is much greater that that of either the skipjack or the herring. The critical age for cod is about 14 years, while that for herring is 3 years and for skipjack tuna less than 1 year. Figure 3.6b shows the effect of sexually dimorphic growth, which is characteristic of some species of fish.

Although specific examples of the behavior of $B(t)$ with changes in W_∞, K, and M are enlightening, a more general picture of the factors that

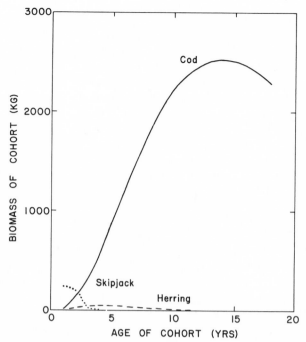

Figure 3.6a Elaboration of biomass in cod ($W_\infty = 24$ kg, $K = 0.1$, $M = 0.1$); skip-jack tuna ($W_\infty = 32$ kg, $K = 0.42$, $M = 1.68$); and herring ($W_\infty = 200$ g, $K = 0.38$, $M = 0.25$).

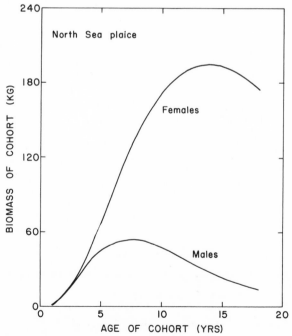

Figure 3.6b Elaboration of biomass of plaice showing sexually dimorphic growth.

control biomass can be inferred from a more detailed study of (3.7). Taking derivatives, we have

$$B' = 3Ke^{-(M+K)t}(1 - e^{-Kt})^2 - Me^{-Mt}(1 - e - Ke)^3, \tag{3.8}$$

$$B'' = M^2e^{-Mt} - 3(M + K)^2e^{-t(M+K)} + 3(M + 2K)^2e^{-t(M+2K)}$$
$$- (M + 3K)^2e^{-t(M+3K)}, \tag{3.9}$$

which, when set equal to zero, gives the age at which the biomass curve reaches its maximum, the age at which the rate of biomass elaboration reaches its maximum, and the age at which the rate of biomass elaboration reaches its minimum for different combinations of K and M (see Alverson and Carney, 1975).

These relations are plotted in Figure 3.7, which shows that the shape of the biomass curves tends to be most variable for any constant ratio M/K and most constant for any constant sum $M + K$. If we assume that the species within a family evolved from a common ancestry and that the ratio M/K tends to be constant within a family, then it would appear that the evolutionary process was such that an important specialization was a change in the position or shape of the biomass curve.

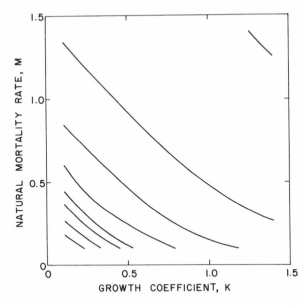

Figure 3.7 Age in years at which cohorts reach a maximal weight for various values of M and K. Note that the age is relatively constant for the sum $M + K$. Age contours are integer-valued and range from 1 *(upper right)* to 8 *(lower left)*. (Developed from computations in Alverson and Carney, 1975: 135.)

Establishing the number of eggs produced. Theory enables us to calculate the biomass of a cohort as a function of its age. Theory also makes it possible to determine the mature-fish biomass in the cohort simply by integrating the biomass function from the age of maturity to some upper limit corresponding to the maximum age of the cohort and the effects of changing growth and mortality on biomass.

Now if the production of eggs is proportional to the biomass, then the biomass of eggs can readily be obtained. But perhaps more important, because the effect of change in growth and mortality on biomass can easily be reckoned, it is also easy to reckon the effects of this change on the production of eggs.

While this approach is a reasonable first approximation to the interrelation of growth and mortality with the production of eggs, we must take additional factors into account. These include the observation that different families or groups of fish partition their energy between gonadal and somatic biomass in different ways. With respect to biomass partitioning, Ware (1975), for example, presents data showing that ovaries in cod and flatfish represent about 20 percent of body weight, while ovaries in herring represent 30 percent of body weight. There seem to be at least two types of partitioning. The first is a temporal partitioning, the second a biomass partitioning. With respect to temporal partitioning, fish become mature at different ages or sizes. Beverton (1963), for example, reports that the length at maturity of herring and anchovy is 65 to 80 percent of maximum length; the length at maturity of cod and flounder is 45 to 65 percent.

Recent work suggests the need for some caution in accepting gonad weight as simply representative of annual egg production. This is because gonad weight usually represents the weight of the ovary at a specific point in time; yet it is becoming more and more evident that some fish spawn batches of eggs over extensive time periods within a year (see, for example, Hislop, 1975).

In addition to the general characteristics regarding egg production, considerable interannual variability can occur in the production of eggs per unit biomass. This interannual variability may be a density-dependent response which would operate when the stock was at a low level of abundance, to reduce the age at maturity, for example, or to increase weight-specific fecundity, or to increase the number of spawning periods in batch-spawning fish. Any of these possibilities would, of course, invalidate the assumption that egg production is proportional to biomass.

Transformation of eggs into recruits: recruitment-stock theory. Stock-and-recruitment theory (primary references are Ricker, 1954; Beverton and

Holt, 1957; see also Cushing's 1981 review) is a major branch of fish-population dynamics that will be discussed in detail in Chapter 5. The phrase *stock-and-recruitment* alters the context of the analysis, because recruitment is usually taken as the dependent variable. Since in the description of a relationship the independent variable usually precedes the dependent variable, we will replace *stock-and-recruitment* with *recruitment-stock*. Recruitment-stock theory portrays recruitment as a function of egg production of the spawning biomass. It is important to note that egg production was the independent variable specified in early studies, but many recent analyses use spawning biomass as the independent variable, thus not taking into account the varying relation of the quantity of eggs produced to the biomass of spawning females or, put another way, implicitly assuming that egg production is directly proportional to mature female biomass, a rather tenuous assumption (see Chapter 6).

Various functional forms have been proposed in the literature to relate the magnitude of recruitment to the magnitude of parent stock, including Ricker's (1954) dome-shaped form and Beverton and Holt's (1957) asymptotic form. Shepherd (1982) has developed an equation that enables us to appreciate the variety of functional forms used to relate recruitment to parent stock:

$$R = \frac{aS}{1 + \left(\dfrac{S}{K}\right)^{\beta}}, \tag{3.10}$$

where R is recruitment, S is stock, and a, β, and K are constants. Scaling this function in terms of $(aK)^{-1}$ and $(K)^{-1}$ gives an asymptotic curve (asymptote with slope zero) when $\beta = 1$. If $\beta < 1$, then recruitment continues to increase with increasing stock. If $\beta > 1$, there is a level of stock at which recruitment has a maximum (Figure 3.8). If (3.10) is a reasonable representation of recruitment and stock, then it is important to determine (a) the values of β for each stock and (b) the mechanisms associated with particular values of β that might be less than, equal to, or greater than unity.

In most analyses of recruitment-stock data, the individual recruitment-stock points do not seem to correspond to the theoretical relation of recruitment to stock. This discrepancy suggests that spawning stock is a poor predictor of recruitment. So although it is relatively easy to predict changes in biomass from changes in growth and mortality, it is more difficult to predict changes in egg production from changes in biomass, and it is quite difficult to predict recruitment from changes in biomass.

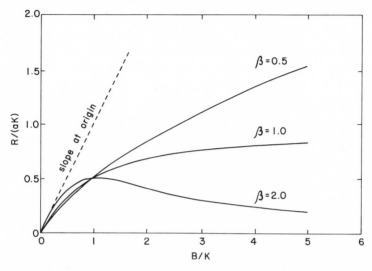

Figure 3.8 Recruitment-stock curves scaled to $R(aK)^{-1}$ and $\beta(K)^{-1}$ after Shepherd (1982: fig. 1). The curves predict recruitment for any magnitude of stock. As β increases, the curve becomes more dome shaped. (Reprinted with permission of ICES.)

The recruitment-stock relation, then, is in this sense the weakest link in the chain uniting the number of recruits, the elaboration of biomass, and the production of eggs with the subsequent number of recruits. This observation has lead some to believe that there is little utility in the theory, and that its study should be suppressed. Yet there *must be some relation* of recruitment to stock; otherwise stocks would not be persistent but would increase exponentially or disappear.

Production-Model Theory

The notion of production is the most fundamental concept in fisheries science. The notion has its roots in the works of Russell and Schaefer, among others. The idea is that increments to the fish population will derive from the numbers of young fish produced each year, while decrements to the population will result from mortality. Obviously, then, the population must be in equilibrium when the increments equal the decrements.

Schaefer (1957) phrased the problem as follows: Let

$$\frac{\Delta P}{\Delta t} = I - D, \tag{3.11}$$

where the change in the population ΔP is the difference between the population increments and the decrements. Let

$$I - D = Kf(\bar{P})\bar{P}, \tag{3.12}$$

which indicates that the net change in the population is proportional, with constant of proportionality K to the product of a function of mean population size and the mean population size. Schaefer specifies that $f(\bar{P})$ be of a form to regulate the population, and the simplest form is that it is the difference between a limiting maximum population magnitude L and the population magnitude (that is, $L - \bar{P}$). Hence we can rewrite (3.11) as

$$\frac{dP}{dt} = I - D = K\bar{P}(L - \bar{P}), \tag{3.13}$$

which is a parabola. In other words, the productivity, or dP/dt, is zero when P is zero, at its maximum when $\bar{P} = \frac{1}{2}L$, and zero when the population is at its upper limit L. Setting $dP/dt = 0$ in (3.15) generates the condition for population equilibrium. If numbers of individuals are removed from the population so that dP/dt is always zero, then the population is always in equilibrium. Therefore expressions of (3.15) with the derivative set equal to zero are called surplus-production functions, because it is possible to remove the surplus production from the population and still maintain $dP/dt = 0$.

The assumption of the form of $f(\bar{P})$ and the specific values of K and L then determine in mathematical terms the productivity of the population. We can see that in circumstances where the productivity of population A, $P(A)$, is greater than the productivity of population B, $P(B)$, for $\bar{P}_B$ and $\bar{P}_A$, then we say that population A is more productive than population B. Yet it is possible that $P(A)$ and $P(B)$ cross for some values of $\bar{P}_A$ and $\bar{P}_B$, and in this case we say that population A is conditionally more productive than population B over the appropriate ranges of $\bar{P}$.

We can see that constant KL refers to *both* growth and recruitment and constant K refers to mortality. Obviously P will be subject to chance fluctuation, but both KL and L could change as a result of biotic or abiotic environmental change. Holding $dP/dt = 0$ then sets constraints on the way that the various constants can vary (see Rothschild and Suda, 1977).

The parabolic model (3.13) was generalized by Pella and Tomlinson (1969) as follows:

$$\frac{dP}{dt} = HP^m - KP, \tag{3.14}$$

where H, K, and m are constants (H has become KL). We can see that setting $m = 2$ in (3.14) yields (3.13). If $m > 2$, surplus production begins

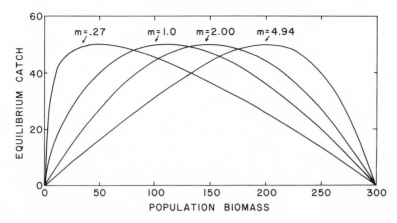

Figure 3.9 Surplus production functions for various values of *m*. (Modified after Pella and Tomlinson, 1969.)

to fall sharply after the maximum. If *m* < 2, surplus production tends to become asymptotic as *m* decreases (see also Fox, 1970). The various forms of the model are shown in Figure 3.9.

Both Shepherd's generalization of the recruitment-stock curve (3.10) and Pella and Tomlinson's (1969) generalization of the production model (3.14) leave us with something of a dilemma, as it is easy to see that generalizations provide the opportunity to obtain a better "fit" to data than could be achieved with more traditional approaches. The problem is that some of the better fits might not make sense. For example, it is not uncommon to obtain best-fit production models with very low values of *m*. Yet such production models are not biologically sensible in that extremely large values of effort will not reduce the catch. Thus the utility of these generalizations involves understanding the as-yet-unknown way in which the parameters change from being constants to being functions of biomass.

A striking feature of all these models is that they address only parts of the population-dynamics process. There is, for example, no explicit theoretical treatment of egg production or the way that the environment affects reproduction, growth, and mortality. Further, the parameters are by definition constants, not functions of time, and theory that does not at least recognize their temporal variability is incomplete.

But no theory is completely general. All we can hope for is to delineate the scope of extant theory and consider ways in which it might be extended. The following chapters examine in more detail the scope of existing models and explore ways of extending them to take greater account of both the stability and the variability of fish stocks.

4

The Context of the Problem

To set the context for our study of the recruitment-variability problem and its implications for fishery management, this chapter articulates the theoretical and practical questions raised by our earlier observations on the mechanisms that induce fish-population stability and variation.

Relation between Fish-Population Dynamics and Fishery-Population-Dynamics Theory

The natural history of marine fish is summarized in Figure 4.1. The figure indicates that a typical marine-fish population might spawn 10^{12} or even 10^{13} eggs (see, for example, Garrod and Knights, 1979). The eggs, about 1 mm in diameter, hatch in a matter of days. The larvae, millimeters in length, have a several-week life span, which is governed by rates of mortality and ontogenetic development. As part of the plankton, the eggs and larvae are subject to water-mass advection as well as to ambient microscale and fine-scale velocities and accelerations. The larvae, leave their planktonic existence as they grow and develop increased swimming capability, becoming juvenile fish at lengths of more than about 15 mm. Remaining as juveniles for a period of a few years, they eventually become mature adults.

The setting for the dynamics in Figure 4.1 is analogous to a chemical-process cross-current extraction system. The vertical flow runs from eggs to larvae, to juveniles, to adults. Horizontal, or cross-current, flows exist at the larval, juvenile, and mature-adult stages. The input to each cross-current flow is ingestion, and the output is associated with predation, natural death, and metabolic losses. The cross-current notation then places each stage within the spectrum of food particles exemplified in

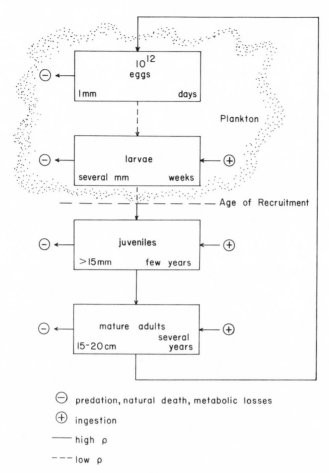

Figure 4.1 Structure of marine-fish life history. Figure shows the transition among life-history stages; some of the statistics pertinent to appreciating the scales of events pertinent to each stage; and the magnitude of correlations between life-history stages (*solid arrows* reflect relatively high correlation; *dashed arrows,* relatively low correlation). The plus and minus signs symbolize the cross-current exchange nature of the stages in the trophic system.

Figure 4.2 (a special case of a spectrum is the size spectrum of organisms in the sea; see Sheldon, Prakash, and Sutcliffe, 1972). This arrangement emphasizes that the dynamics of each stage depends on the dynamics of the organisms that the population ingests as well as those of the organisms that are its predators.

The dynamics implied by Figure 4.1 deserves special attention. The

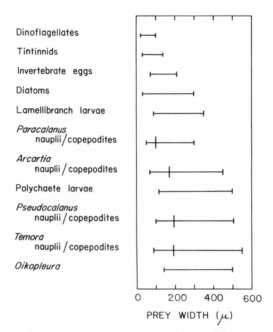

Figure 4.2 Size range of larval foods in the North Sea. Size range of fish larvae is about 400 to 1000. Smaller particles than the food include small protozoa, flagellates, diatoms, and bacteria. Some zooplankton are larger than some ichthyoplankton; hence the mix of food items for fish depends on cross-current flow from bacteria to large zooplankton. (Data from Last, 1978a: fig. 3; 1978b: fig. 3.)

annual production of a very large number of eggs by each population is a dominant feature. To appreciate the magnitude consider placing the 10^{12} eggs end to end. The line of eggs, each 1 mm in diameter, would be 10^{12} mm, or 10^6 km, in length (the earth's circumference is only 4×10^4 km). Since the magnitude of the adult population is much less than the number of eggs, a substantial mortality must be incurred by the eggs and larvae. The great diminution of eggs and larvae provides considerable opportunity for large variability in mortality, which seems to stabilize near the terminal portion of the larval stage or in the early juvenile stage (see, for example, Cushing and Harris, 1973: 142–144). Accordingly, the stabilization age partitions the fish's life into a planktonic phase, where the correlation between the numbers of older individuals and their young progeny is relatively low, and the juvenile-adult-egg phases, where the correlation between the numbers of juveniles and adults and between the numbers of adults and eggs is relatively high.

An important feature of the dynamics is that the transfer of material

among life-history stages is not likely to be represented by a direct proportion or percentage. Generally speaking, trophodynamic or other
transfers, although often presented as simple proportional relationships,
should have a curvilinear relation of input to output; hence any single
estimate of transfer efficiency — 15 percent, say — must be biased (note
in Figure 1.6, for example that the energy-transfer coefficients are constant). Curvilinear density-dependent effects reduce growth and increase mortality at high population levels, and vice versa.

The dynamics in Figure 4.1 also reflects the amplification of energy
content from one stage to the next. This condition is exemplified by the
amplification of the variability associated with energetically minute
"parcels," the eggs and larvae, into the variability of energetically large
parcels, the adult fish (Figure 4.3). It must be remembered that amplification may affect the variance of the biomass as well as its mean value.
The amplified signal contains components of the somewhat abstract
microscale and fine-scale ecosystem, which is driven by, and embedded
in, physical and biological events that operate at larger scales.

The mechanics transferring individuals or material from one life-
history stage to the next is shown in Figure 4.4. The cohort is initialized by
a particular number of young fish, or recruits. Mortality, taking its toll,
continually diminishes the number of recruits. At the same time, each

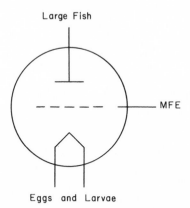

Figure 4.3 The amplifier effect. In a vacuum-tube amplifier a large flow of elec-
trons passes from the cathode to the anode. The flow is modulated by a small
voltage applied to the grid. Hence the variability in the small voltage applied to
the grid is amplified at the anode. By analogy, variability in the microscale and
fine-scale ecosystem (MFE) is amplified in the transition from eggs and larvae to
large fish.

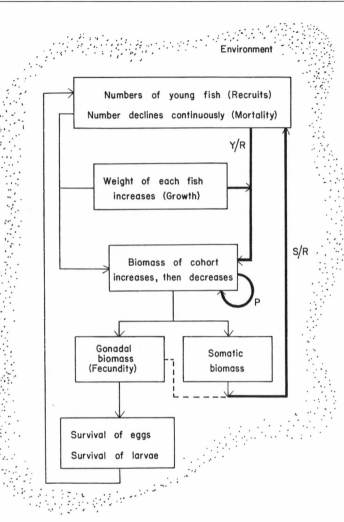

Figure 4.4 The mechanics of transfer among life-history stages is represented by *light lines*. Theories that attempt to account for these transfers are the yield-per-recruit theory (*Y/R*), the production theory (*P*), and the recruitment-stock theory (*S/R*), represented by *heavy lines*. The *dashed line* connecting reproductive biomass and somatic biomass reflects that recruitment-stock relations are presented as functions of either reproductive biomass (which is consonant with the available theory) or somatic biomass (in actuality, somatic biomass plus some usually unknown quantity of reproductive biomass).

living recruit continues to grow, increasing its individual weight. The cohort as a whole likewise increases its weight during its early life because integrated growth is greater than integrated mortality. At some point a maximum is reached, after which the cohort declines in weight.

The partitioning of somatic and reproductive biomass in Figure 4.4 draws attention to the inappropriateness of treating growth as a process that reflects a homogeneous or simultaneous change in all organ systems. Before maturity, for example, growth involves the elaboration of essentially somatic tissue; after maturity, growth involves the elaboration of both somatic and reproductive tissue, the reproductive tissue generating gametes that sequentially become zygotes, larvae, juveniles, and recruits.

Surprisingly, only a few of the transitions shown in the figure have been addressed by theory: yield-per-recruit theory depicts cohort biomass as a function of age (time) by integrating the product of the initial number of fish in a cohort and their mortality and growth rates; surplus-production theory takes into account the relation of the biomass rate of change to the biomass; and recruitment-stock theory (usually using total biomass as a proxy for reproductive biomass) expresses the number of recruits as a simple function of mature-fish biomass. Although the relation of recruitment to stock is our ultimate theoretical concern, note that the theory addresses only the direct relation of reproductive or somatic biomass to recruitment; it does *not* explicitly take into account specific intermediate steps, such as egg or larval survival. Further, the veracity of the theory has been difficult to assess, as it concerns the relation of egg production to recruitment; yet many of the reports in the literature purporting to represent recruitment-stock theory present data on the relation of recruits to the biomass of spawning fish. Thus, as Figure 4.4 shows, important parts of the population-dynamics process are not represented by standard theory, namely, (a) the partitioning of mature-fish biomass into somatic and reproductive components, (b) the relation of the number of eggs spawned to the number of young fish that survive to maturity, and (c) perhaps most obvious, if not most important, the relation of the physical environment to any part of the population-dynamics system (with the possible exception of the relation of temperature to growth).

Biological Information in Fishery Management: Traditional Theory and Modern Problems

Fishery management regulates fishing intensity to obtain a harvest that is optimal in some sense. Steps involved in fishery management include collecting basic biological, economic, or societal data on the fishery; transforming the data into information useful for management decision making; and making and executing management decisions.

The discussion in this book concentrates on the fishery manager's requirements for biological data and processed information. These information needs may have either a traditional or a modern orientation. The traditional orientation is associated with traditional fishery models: the yield-per-recruit, stock-and-recruitment, and production models discussed earlier. In contrast, the modern orientation is more concerned with the development of new models and techniques for predicting year-to-year recruitment variation and understanding its causality, and for managing habitats. Questions framed in the modern orientation are more relevant to broader societal issues and are considerably more complex, more difficult, and more demanding of data than those of the traditional bent. The broader issues involve annual variability and its causality, ecosystem management, stock-rebuilding strategies, habitat management, and aquaculture.

Traditional Orientation

The traditional models have considerable utility because they can be used to determine optimal levels of average yield per recruit (or stock biomass), average recruitment and average surplus production. But with regard to our major interest, prediction, each of these models has the same limitation. Because the number of variables in the system is much larger than the number of parameters in the model, the error structure in the model contains more than chance deviations from the model parameters; it also includes variability associated with the variables extrinsic to the model. The utility of the traditional model for gaining an insight into the system or for prediction depends on the extent to which the future error structure replicates the past error structure. Increased sample size normally improves statistical precision, but larger sample sizes suggest a greater likelihood that the future error structure will change substantially.

Put another way, a change in the error structure is tantamount to a change in the *stationarity* of environmental *conditions*. The conditions may relate to biotic factors associated with the stock of concern, to biotic factors associated with other biota, or to physical features of the environment. When conditions change such that the twenty-to-fifty-year time series can no longer be considered representative of either the important future variables or their effects, there is then a concern for the validity of the traditional model. Hence traditional models do not take into account changes in environmental conditions external to the model. That this is a problem is reflected in the occurrence of events that cannot be predicted by traditional models, such as stock collapses and explosions. Understanding these environmental conditions is a major problem in fishery management and in population dynamics as well.

What are the conditions, and how can we determine whether they are stationary or stable? This question is extremely difficult because, in the first place, the precise definition of "environment" is somewhat abstract. Even if the variables were well identified, it would be difficult, without additional information, to determine whether the conditions are stable, because by definition the stability assertion implies the availability of information external to the data set at hand. Thus stock "collapses" or "explosions," for example, could be taken to imply simply that the magnitude of the collapse or explosion is greater than the range of variability observed up to the time of the unusual event (see Figure 2.1). The collapse or explosion could reflect change in environmental conditions, thus highlighting a further distinction between traditional and modern approaches—namely, that traditional approaches attempt to predict stock productivity under circumstances where conditions are *assumed* not to have changed, whereas modern approaches attempt to predict the effects of changes in the conditions. Hence the utility of traditional models is limited by their exclusion of environmental theory and their dependence on time-series data.

Validity of particular models. Aside from these general considerations, it is important to identify problems associated with particular models. Procedures for estimating yield-per-recruit functions, for example, seem straightforward because they do not make explicit use of time-series information, and variations in recruitment are assumed not to affect the yield per recruit. This belies the considerable complexity underlying biomass or yield-per-recruit estimation, as the yield-per-recruit function is itself dependent on estimates of natural mortality and growth that are normally derived from time-series data (in any event, it is difficult to be confident of estimates of natural mortality, let alone interannual changes in natural mortality). In addition, yield-per-recruit models also require independence among the variables associated with growth, mortality, and recruitment magnitude. It is conceivable that some management strategies suggested by yield-per-recruit theory would be valid only over a certain range of recruitment values, and conversely certain yield-per-recruit strategies would not be employed if it were thought that they would affect recruitment. In other words, use of yield-per-recruit models assumes that there is no density-dependent growth or mortality.

The validity of the yield-per-recruit model requires stability of growth and mortality, and the assumption that recruitment is not related to yield per recruit or to stock biomass needs to be maintained. Hence the validity of yield-per-recruit or cohort biomass models requires external information on stability of mortality, growth, and recruitment. In particular, serious use of the yield-per-recruit model requires addressing the fundamental nature of the recruitment process to ascertain the interactions among recruitment, natural mortality, and growth.

Surplus-production models also present certain problems. A key problem involves the functional form, or "shape," of the surplus-production-model curve. According to theory, it is possible to obtain a relation of surplus production to stock size that ranges in form from a traditional parabola to an exponential or asymptotic form, even including a "backward-bending" function (see Clark, 1974, and Gulland, 1977). The true form of the underlying function can be critically important in fishery management, where it is important to know, for example, whether the function is parabolic rather than asymptotic. If it is parabolic, additional units of fishing mortality will cause the catch to decline, whereas if it is asymptotic, additional units of effort will result in a stable catch. Unfortunately the differences among the various forms of the curves are evident only at high levels of fishing mortality. This fact is a serious concern in fisheries management, where forecasts of effects of high levels of fishing mortality are often required even though such levels have not been experienced. Hence, for management purposes assumptions must be made about the form of the curve, or additional theoretical information must be developed on the dynamic behavior of the stocks at high levels of fishing mortality, particularly with respect to recruitment.

Although the asymptoticlike form of the surplus-production model does not seem logically conceivable, it commonly occurs in practice. If the asymptotic form is not a result of statistical error, then determination of (a) the conditions that produce the asymptote and (b) the conditions that will eventually cause surplus production to decline, as it must, is a major population-dynamics problem.

Examples exist where production-model conditions have apparently changed, as when a relatively large year class or series of year classes moves through the fishery (see Rothschild, 1971). Possible examples of changing conditions may have occurred in the eastern tropical Pacific fishery for yellowfin tuna. Early studies of the fishery (1950s) reported that fishing was conducted by bait boats operating close to the coasts of North and South America. At that time there was a general belief that yellowfin tuna were not common between these coastal areas and areas west of the longitude of Christmas Island. Estimates of the surplus-pro-duction-model parameters suggested that the fishery could sustain catches of roughly 100,000 tons (Schaefer, 1954, 1957). Subsequent de-velopments in the fishery involved technological improvements and a shift from the less efficient bait-boat gear to the more efficient purse-seine techniques. At the same time the fishery tended to move offshore into areas where it was formerly thought that large yellowfin tuna did not exist. Substantial quantities of the fish were caught in these areas, and recomputations reflected that the fishery could sustain yields of well over 200,000 tons. There is little doubt that Schaefer's calculations reflected the sustainable yield of the stock as it was defined at that time, but what is not clear is whether the total stock was capable of yielding only about 100,000 tons and later increased its productive potential, possibly be-cause of environmental change or increased food supply resulting from increased fishing for other species of tuna in the same waters, or whether the stock was always capable of yields of 250,000 tons each year.

Again, insights into the form of the production model cannot be ob-tained from the model itself. Additional information is required, largely on the variables associated with recruitment. Recruitment variability must be important in determining the form of the production function, the obvious effect of periodicity in large and small year classes on surplus production, and responses in the levels of production to environmental change.

Finally, recruitment-stock models are the centerpiece of fisheries theory in terms of both its significance to the population-dynamics pro-cess and its practical importance. Recruitment-stock theory provides the manager with insights into the most effective management approaches. For example, if stock is a good predictor of recruitment, then the manager can manipulate stock to control recruitment to some desired end. If stock is a poor predictor of recruitment, then other approaches are necessary, because under such circumstances the manager can only take into ac-count expected variability and use this information directly or devise variability-reduction strategies.

As two-dimensional representations of a phenomenon of much higher and shifting dimensionality, recruitment-stock models are extremely simplistic. Their effective use requires a substantial amount of informa-tion extrinsic to the theory. There are many situations, much as in the production-model case, where it is important to distinguish between various functional forms (for example, Figure 3.8). Yet the precise shape of the curve cannot be determined without additional information on the underlying nature of the recruitment process.

Interpretation of the recruitment-stock relationship gives rise to practical management problems. For example, in some fisheries it appears that there is "no stock-recruitment relationship" that is, the recruitment-stock relation is essentially a straight line with slope zero. The assertion is often taken to mean that there is no need to monitor fishing effort, because fishing (or, equivalently, stock size) does not affect recruitment. Actually, under these circumstances there is even more reason to monitor recruitment. There must be a recruitment-stock relation, in the sense that the recruitment-stock function must pass through the origin, or close to it. In stocks with an empirical zero-slope recruitment-stock relation, the approach to the origin must be quite precipitous; hence it is important in management to determine the precise position of the point at which recruitment would sharply decline. The determination of this point cannot generally be determined from past information, so recruitment-stock theory needs to be supplemented with additional information.

The question of overfishing. What is overfishing? Each of the three models contains explicit or implicit information on overfishing, a traditional concern of fishery managers. Each model exists in two forms: one with a maximum and one without a maximum, a form that is essentially asymptotic. Clearly levels of fishing mortality that will produce optimal yield per recruit, recruitment, or surplus production can be determined only from functional forms possessing maxima. Levels of fishing mortality in excess of the maximum result in overfishing, and levels less than the maximum result in underfishing. Biological overfishing, and to some extent underfishing, cannot be determined from asymptotic forms of the models, where there is no apparent negative biological effect from the increase in fishing mortality — that is, there are no levels of fishing mortality such that increases in mortality result in decreases in yield per recruit, recruitment, or production. (The application of economic criteria will often generate an optimum in the asymptotic biological models. The requirement of a maximum in one of the population-dynamics functions has been obviated by the so-called $F_{0.10}$ criterion [Gulland and Boerema, 1973], which can easily generate diseconomic management advice.) Stocks have been thought to be "stock overfished" if the magnitude of size-specific fishing mortality exceeded the level that would produce the maximum yield per recruit, and "recruitment overfished" if fishing intensity exceeded the level that would produce maximum recruitment or where recruitment "collapsed" under apparently heavy fishing (see Cushing, 1977).

In summary, the validity of the three traditional models is challenged when it is necessary to extrapolate either in time or beyond the range of extant data. For example, a prediction of either yield per recruit, stock and recruitment, or surplus production for the twenty-first year based on twenty years of past data might be required. If the underlying conditions have not changed—that is, if the statistics relevant to the parameters to be estimated for the twenty-first year are drawn from the same error structure of statistical distribution as that which existed for the first twenty years—then the task is relatively straightforward. If conditions have changed, then supplemental information on the conditions is required.

In addition to extrapolation in time, extrapolation of the independent variable—fishing mortality or stock magnitude—generates concerns regarding the functional form of the model. For example, estimates of traditional-model parameters based on twenty years of observations may be derived from values of fishing mortality that have not exceeded $F = 1.5$. But what happens when $F = 2.0$? Is the underlying functional form really different from that which has been assumed, considering that the differences are detectable only at high, usually unexperienced, levels of fishing mortality?

From a fishery-management point of view, these distinctions are not fine points. These biological concerns have important economic implications because, while much fishery management, at least in principle, depends primarily on economic rather than biological criteria, the economic models generally depend on the biological models, and any concern for the validity of the economic models must certainly reflect a deeper concern for the validity of the biological models. In fact, the greatest economic impacts are likely to derive from the unexpected, a sudden departure from the average events predicted by traditional biological models. Neither continued use of traditional models nor additional data will ameliorate these problems.

Modern Orientation

The modern approach is an *extension* of the traditional approach and is not necessarily "better." Modern concerns are more demanding of both data and concepts; they no longer focus on "on-the-average" events but examine events that occur in particular years. They thereby diminish the importance of averaged temporal observations of stock, recruitment, or environment, and they question the usefulness of past observations to gain explanatory insight into variability. Beyond fishery management, in the traditional or narrow sense, modern concerns involve ecosystem management, stock-rebuilding strategies, habitat management, and aquaculture.

Ecosystem management. A heightened awareness of resource management spurred by excessive fishing by distant-water fleets and extension of coastal-state fishery jurisdiction in the 1970s focused attention on the traditional models, which were roundly criticized as being single-species, deterministic, biologically oriented models that were unresponsive to management requirements for information on multiple-species interactions, year-to-year variability, and economic and social values (Larkin, 1977; Holt and Talbot, 1978). To correct some of these faults a management of the entire ecosystem was proposed to replace the traditional approach. But implementation of ecosystem management was stymied by substantial though predictable difficulty. For example, before an ecosystem-management information base could be developed, it was first necessary to develop a multiple-species theory and a body of empirical data to test and modify the theory. It was relatively easy to generalize mathematically the single-species models into multiple-species formats, but exceedingly hard to estimate the coefficients pertaining to species interactions (see, for example, FAO, 1978; Pauly and Murphy, 1982; Mercer, 1982). The complex task of predicting recruitment is a prominent stumbling block in the formation of a multiple-species theory and the attainment of even a rudimentary ecosystem-management strategy.

If the relation of recruitment to stock is not understood in a single-species context, how can it be understood in a multiple-species context? The question has two partial answers, which are contradictory. The first is that the single-species context is simpler than the multiple-species context. Therefore understanding stock and recruitment in the single-species context is a prerequisite to understanding it in the multiple-species context. The second is that other species may be a major factor affecting the single-species recruitment-stock relationship, so difficulties arise in understanding the single-species problem because multiple-species relationships are ordinarily not taken into account. In any event, since recruitment drives single species variability, it also drives multiple-species variability, and any advances in ecosystem management must be preceded by an improved understanding of the recruitment problem.

Stock-rebuilding strategies. There have been several instances of so-called stock collapse. The question is raised as to whether the collapse was caused by fishing, by natural phenomena, or by a combination. It is commonly believed that modern stock collapses are caused by fishing, but there is evidence that other factors are also involved. The key is the relation of fishing mortality to stock size. If the stock declines without an increase in fishing mortality, then factors other than fishing must have triggered the collapse.

Jakobsson (1985), for example, has presented time-series data on fishing mortality and stock abundance for several herring stocks. Figure 4.5a

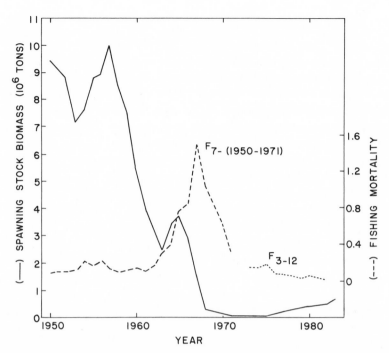

Figure 4.5a Spawning-stock biomass (Icelandic summer spawners) and fishing mortality rate (*F*) for 3-year-old and older herring, 1950–1982. (From Jakobsson, 1985: fig. 16.)

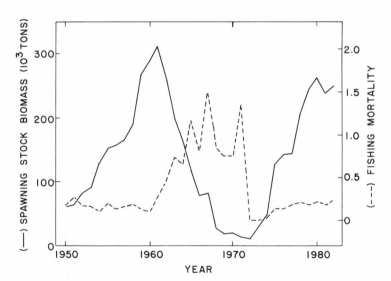

Figure 4.5b Spawning-stock biomass (1950–1982, Norwegian spring spawners); fishing mortality (*F*) for 7-year-old and older herring (1950–1970); and fishing mortality (*F*) for 3- to 12-year-old herring (1973–1982). (From Jakobsson, 1985: fig. 5.)

shows his time-series graph for Icelandic summer-spawning herring; Figure 4.5b, Norwegian spring-spawning herring. Icelandic summer-spawning stock increased under fairly constant fishing mortality until about 1960. As fishing mortality increased, the stock began to decline. Later, as fishing mortality decreased, the stock increased again. In other words, fluctuations in icelandic summer-spawning herring-stock abundance appear to be closely coupled with fishing intensity. The Norwegian spring-spawning herring, however, provides a contrasting example. This stock began its decline before any great intensification of fishing effort occurred and has not recovered even though fish mortality has declined. Thus fluctuations in Norwegian spring-spawning herring do not seem to be closely coupled to fishing mortality, at least in terms of this simplistic interpretation of Figure 4.5.

If fishing causes a collapse, then we might surmise that a moratorium on fishing would cause the stock to increase to its former level of abundance. But as it is generally not known whether declines in recruitment are caused by fishing or by environmental changes, or by some combination of the two, it is not immediately obvious whether a fishing moratorium would result in an increase in stock. Observations of stocks that have experienced drastic declines in abundance reflect that, whereas in some instances a cessation or reduction of fishing is associated with the stock increasing to its former level of abundance, in other instances substantial reductions in fishing effort have not been accompanied by a recovery in abundance (as yet).

The best strategy for managing a collapsed stock depends on whether the stock has an inherent capability of increasing, or whether fishing or other causes have for some reason generated a permanent or semipermanent depression in recruitment success. If the stock has an inherent capability of increasing, then the costs and benefits of a moratorium on fishing could be evaluated. But if the stock has lost this inherent capability, it would be necessary to adjust long-run fishing effort to a much reduced stock level. Thus the success or failure of a stock-rebuilding strategy depends to a large extent on whether the stock has retained or lost its inherent capability of increasing to precollapse levels. This observation again focuses on the recruitment problem and the identification of the conditions associated with "good" recruitment and "poor" recruitment.

Habitat management. Habitat management encompasses a wide spectrum of environmental decision making. Much is at stake, because these decisions hold economic and social development in balance with presumed environmental degradation. Although fishery management has tended to be restricted to "fishing" in the narrow sense, the economic and social stakes associated with fishery-related aspects of habitat management are often much broader.

Habitat-management issues involve, for example, decisions affecting activities such as power-plant siting, harbor construction or modification, mineral-extraction activities, and the management of industrial or sewage effluents containing toxic or eutrophicating substances, as well as nonpoint source pollution. In addition to particular siting activities, much more general concerns exist for the destruction of mangroves, nursery areas, and estuarine habitats as well as for the environmental impact of industrialization, such as the formation of acid rain, or for the impact of agriculture, such as eutrophication.

The problem is that the effects of these human activities are appreciated only when they operate at their most intense levels; hence judgment on what is desirable or undesirable derives more from public opinion based on short-term observations of unusual extremes than from technical assessment of the effects of specific habitat-modifying activities. Lack of sound technical analysis suggests that some popular environmental decisions are harmful; that society's limited energies may be devoted to less important environmental problems; and that the average person thinks that ecosystem dynamics is well understood, an erroneous belief promoting a false sense of security and stifling the development of improved environmental information.

In fact, insufficient information exists on the population-dynamics effects of environmentally ambient concentrations of contaminating or eutrophicating agents. To be sure, studies of the effects of pollutants where experimental concentrations far exceed the typical exposures of feral animals are numerous. Also commonplace are studies of the effects of lesser concentrations of a pollutant — so-called sublethal studies — reporting the "body burden" of a toxic substance. But studies that consider the true effects of a toxicant or habitat modification — the changes in population dynamics that result from possible changes in mortality, growth, or recruitment — are generally missing from the research portfolio.

Of course, concern is not for just the population dynamics or the changes in temporal abundance of a single species; rather, it is for the population dynamics of all the organisms in the ecosystem as a whole. The complications are obvious. Introduce a toxicant, and suppose it has a negative physiological effect on species A and no negative effect on associated species B. Suppose species A is essential food for species B. Then the toxicant reduces both species A and B, although it might not be observable in B. Furthermore, if species A is perceived as economically unimportant and the effect exists in its dynamics, not in its body burden, then it is unlikely that the problem would even be considered or noticed. It is also conceivable that reductions in species A would not necessarily have a negative result on all associated species; in some instances a decline in species A could cause a sharp increase in associated species, say C and D.

Such complications generate a problem of considerable concern and even greater subtlety. If population dynamics provides the most appropriate metric for measuring anthropogenic effects but is scarcely used to assess anthropogenic change, then is it not possible that the most serious anthropogenic effects are not even known? Is it not possible that some environmental-impact studies give blessing to very harmful projects, while others deny projects that are possibly much less harmful?

What seems to emerge is that society needs to focus its attention not so much on the quantity of particular substances introduced in the environment, not so much on the body burden of these substances, but rather on the direct and indirect effects of substances in various time-space concentrations on the population dynamics of the organisms. In other words, to what extent will the introduction of a toxic substance or the physical modification of a habitat affect the temporal trends in abundance of the species in the ecosystem? In fish, the study of this question must begin with the study of recruitment mechanisms and the conditions associated with recruitment.

Aquaculture. Aquaculture is the artificial propagation of fish, shellfish, and certain algae. Several aquacultural problems require an understanding of the recruitment process. These problems involve the acquisition of brood stock, the propagation of brood stock, and the construction of hatcheries. With regard to brood-stock acquisition, there are several aquacultural endeavors where young feral fish or shrimp are acquired and then reared to marketable size. Variations in recruitment affect variations in the availability of brood stock; hence it is important to know and predict the causes of these variations. In addition, the capture of very young fish is in fact a fishery, so it is important to know the long-term effects of the young-fish fishery on the reproductive potential of the population. Fisheries for very young fish are unusual; while some are targeted on brood stock for aquacultural purposes, others capture very young or postlarval fish for human consumption. For example, there are established fisheries for whitebait; for very young European eel; and, in Japan, for *shirasu*, which are primarily postlarval sardines, round herring, and anchovies (Hayashi, 1961). Detailed study of young-fish fisheries would provide some interesting opportunities to study the effects of different levels of predation (fishing) on the growth and mortality of the cohort during the time interval before it reaches the conventional recruitment age.

In other areas, such as salmon and oyster aquaculture, brood stock is actually propagated. Because the primary intent of propagation is to obtain the most valuable return of adults, it is important to know how the magnitude of this artificial recruitment affects the stock. Obviously at some point production of additional units of brood stock becomes inefficient, and the biological aspects of this problem involve questions of stock and recruitment (as well as of the genetic integrity of the stock). Again, there are important research opportunities utilizing the ability to control recruitment in order to study the effects of recruitment on parent stock.

Finally, since the end of the nineteenth century, hatcheries have been constructed with the idea that the release of young fish would increase the abundance of the adult population. In some instances this scheme has worked, most notably in salmonid culture. Yet in many instances it has not worked. Since hatchery programs involve large public-sector investments in capital and labor, it is important to determine whether any particular hatchery scheme is feasible. Hatchery schemes fail because sufficient hatchery fish do not survive to the stage where they produce an economic or societal return commensurate with the cost of production. The failure of these fish to survive is due to unfavorable environmental conditions (or to a change in the genetic constituency of the fish). These conditions are not generally known, and so their determination is an important research goal.

In sum, fishery management has concerns that are both traditional and modern. The traditional concerns are well defined; thus the boundary separating what is known and what is not known is quite distinct and can be advanced through a better understanding of the recruitment process. The modern concerns are less well defined in the sense that, from a comparative point of view, theory is less well developed than in the traditional models. It is evident, however, that the development of modern theory will also be centered around the recruitment process. Fishery-management applications clearly require an enhanced, modified, or restructured theory that explains the mechanisms that induce recruitment variability in fish stocks. A first step in building the theory is to examine the extant fundamental knowledge to identify those elements that can contribute to revising our understanding of the recruitment process.

Fundamental Knowledge, Biological Oceanography, and Population Dynamics

The concerns associated with biological aspects of fishery management depend heavily on understanding recruitment variability, which in turn requires an understanding of the fundamental processes that drive the variability. The disciplines explored here may give helpful insights into these processes.

The choice of disciplines is rather arbitrary, owing in large part to the hierarchical nature of disciplinary knowledge (for example, both biological oceanography and population dynamics could be considered subdisciplines of ecology). Borrowing from the notions of system analysis (as in Churchman, 1968), it is important to define disciplines iteratively and in a way that is neither too aggregative nor too narrow, so that important elements are not lost in overgenerality or excluded from consideration.

The initial selection of disciplines is based on an awareness that the population-dynamics process operates at both temporally localized and evolutionary time scales. One approach to resolving this complex of temporal events is to consider the temporally local events and then study the mechanisms of its evolution. Adopting this strategy, as a first iteration, we focus on the disciplines of biological oceanography and population dynamics rather than population biology and genetics.

Biological oceanography and population dynamics are similar in that they both consider the formation, transformation, and dissolution of biomass; but they differ in an important way. Studies of biological oceanography, or biological productivity of the sea, are generally concerned with the *bulk* transfer of energy or material, whereas studies of population dynamics are concerned with the transfer of various-sized packages of energy or material, where the packages are individual organisms. Certain aspects of ocean-productivity and population-dynamics studies could strengthen the bridge between these disciplines and the study of recruitment variability.

Biological Oceanography and Ocean Productivity

The study of ocean productivity embraces a wide range of topics. To facilitate its analysis I shall partition the subject into three hierarchical levels. At the highest level are studies of the entire ecosystem, which generally specify the magnitude of primary production, the magnitude of some higher-order production, and the pathways and fluxes of energy or material that interconnect primary production with higher-order production (see, for example, Steele, 1974). These studies are necessarily simplistic because of the practical need to represent the infinitude of entities, compartments, or states and their interconnections in as frugal a manner as possible. At the next level are studies restricted to major components of the productivity system, such as the primary or secondary production of a particular ecosystem. Primary-production studies are usually well bounded and can often be considered independently from the study of other ecosystem components. The study of elements of higher-level production is more problematic because it is difficult to provide a unique view of the complex system. (Many fish, for example, have life histories that are woven through the entire production process owing to their larval, juvenile, and adult existence.) Finally, at the most subordinate level numerous specialized studies consider the ecology, physiology, or behavior of the flora and fauna of the sea.

Models of production at the highest level. Our primary concern is at the highest level of the hierarchy, because perceptions here shape our views on the relative importance of components of the production system and their interconnections. These perceptions in fact serve to allocate emphasis to particular subject-area studies at the subsidiary hierarchical levels (for example, Pomeroy's 1974 paper contributed to the shift of interest to picoplankton and nannoplankton). If perceptions of the system as a whole are not well formed, then studies conducted at subsidiary levels may neglect the most important problems and hence will not contribute effectively to a holistic understanding of the system.

Our concern for fish-population dynamics requires an understanding of the entire system because the energetic exchanges associated with fish production are not restricted to one part of the ecosystem model but exist at various levels of the system. Jones (1984) appreciated this relationship in his comparison of primary production and fish production of the North Sea and Georges Bank. He noted that indices of fish production and primary production were sufficiently proximate, taking into account the dissipation of energy between primary production and fish production, to challenge the assumption of a 10 percent transfer coefficient between trophic levels. He pointed out, for example, that estimates of food consumption accounted for roughly 5 percent of the total particulate carbon fixed each year. Hence with an energy-transfer coefficient of 10 percent "there could be only 1–2 steps between primary production and fish food consumption," which seemed to Jones to be rather low, given an understanding of the piscine food web. He thus concluded that primary production was underestimated, or the 10 percent transfer coefficient was too small. Jones argued that the more likely solution might lie in revaluation of transfer coefficients to take into account the possible winter dormancy of some fish, as well as the fact that small fish evidently use energy more efficiently than large fish. Jones's study and the studies that he cites (for example, Steele, 1974; Sissenwine et al., 1982) reflect the wide range of intepretations regarding the magnitude of primary production and fish production (or production at higher trophic levels), and the pathways between the two levels of production.

Solutions to the problems raised by Jones (1984) are not immediately evident, beyond general observations on model structure and associated parameter estimates. The concern for structure is obviously more fundamental than the concern for parameter estimates because structure defines which specific parameters need to be estimated. A first glance at the problem of structure might suggest that the models described by Jones are too simplistic. After all, any resemblance between the complexities of the oceanic ecosystem and a handful of boxes, interconnecting arrows, estimates of the amount of carbon in each box, and the flux among boxes would be surprising. More complex models might seem appropriate. Increased complexity, however, may be interpreted in terms of increasing model size or linearly manipulating the position of the boxes, rather than in terms of actually modifying the structure of the model. Experience has shown that increasing model size results in larger numbers of permutations and combinations of initial conditions and equations, increasing model "bookkeeping capability" and perhaps reducing intellectual control. Larger models in themselves do not necessarily reflect an increased understanding of the system, because models do not create understanding; they simply record whatever has been understood.

Departures from simple energy-flow diagrams are becoming more evident. In a recent symposium (Ulanowicz and Platt, 1985) several papers were presented that reflected a radiation of approaches ranging from thermodynamics, statistical mechanics, ataxonomic aggregation (primarily size structure), and flow analysis to information theory. The approaches might be divided into two categories. In one category the authors have attempted to apply particular physical principles to ecology; while these principles have an overarching applicability in that all systems behave according to the laws of thermodynamics or gravity, for example, it is not always clear how one might dissect from the complex ecological system those isomorphisms that are completely circumscribed by simple application of physical laws. In the other category the authors have attempted to transfer well-known analytical techniques from other fields (such as input-output analysis) to the analysis of ecosystems. All these technique-oriented procedures are logically equivalent in that they transform components of real-world data into mathematical language. In this sense the use of input-output analysis is no different from the use of differential equations.

The basic questions are whether one technique is better than another, whether more techniques should be explored, and how to choose among alternative techniques. The most common, although generally implicit, criteria are those of insight and predictability. Insight is, of course, very personal and subjective, and predictability has seldom been achieved in complex natural systems. These newly applied techniques for ecosystem analysis may be exciting, but they do not seem to penetrate the two fundamental characteristics of real-world ecosystems: (1) the density-dependent reproductive processes in the dynamic stability-inducing regulation of natural systems, and (2) the complexity of these systems, requiring an explicit set of rules to reduce system complexity to any mathematical or conceptual framework. In short, what is needed is a better understanding of dynamic regulation in the ecosystem and a better understanding of how to analyze complex systems.

Development of alternative structures. Advancement of our understanding of ocean productivity requires emphasis on the development of alternative structures rather than simply constructing larger models or developing improved estimates of the parameters. The development of alternative structures requires a logical framework. The need for such a framework can be appreciated by considering the relation of the real world to holistic models intended to represent the real world. The real world consists of an infinitude of entities and connections or pathways among them. It follows that an infinitude of different models could be used to represent the real world; yet the literature contains relatively few. Thus it is important to consider the logical framework, the syntax, and the criteria that allowed these models to be chosen from among myriad possible models. The existing syntax appears to be based primarily on traditional fishery models and food-chain or food-web analysis. While such an analysis provides interesting insights into the magnitude of energy or material in various compartments of the model, it does not address major components of the system—partitioning of reproductive and somatic biomass, specific events linking spawning and recruitment, and environmental theory—or well-known system dynamics. Models intended to parallel the real world must take into account amplifier and density-dependent effects.

A revised syntax for the development of productivity models requires a better understanding not only of structural questions such as amplifier and density-dependent effects but also of the boxes or compartments constituting the model. For example, Platt, Lewis, and Geider (1984), Goldman (1984), and Gieskes and Kraay (1984) outline serveral concerns regarding the specification of primary production, chiefly in the oligotrophic sea (see McCarthy's 1984 review). There is a similar concern with respect to secondary production, even though most discussions on secondary production skirt the structural issue (see, for example, Williams, 1984).

But even if specification of primary production at a space-time point were resolved, there are a number of issues reflecting problems of regional and annual measurements. For example, a recent revaluation of annual primary production in the Georges Bank area resulted in *doubling* the traditional estimates of primary production for that area from 150 $gCm^{-2}y^{-1}$ to about 300 $gCm^{-2}y^{-1}$ (Cohen et al., 1982). The reassessment of production is evidently due to an improved understanding of variability in intra-annual production. The transfer of primary production *as a unit* to larger herbivores must take into account intra-annual production, not only in terms of its total quantity but in terms of its species composition as well. Because intra-annual variability in production can be large, average annual production values raise questions as to the representativeness of certain integrations or averages of production. Some citations in the literature, for example, allege that primary production varies by a factor of 3. This range of variation has been compared to the typically tenfold range of variation in particular fish stocks, suggesting limited coupling between primary production and fish stocks because fish stocks are more variable than primary production. Inferences in this matter need to be made cautiously because among other things, primary-production variability averages the variability of many species taken together, while fish-stock variability is that of a single species.

In fact, the significance of any particular measurement of primary production in the context of the entire productive system needs considerable study, because evidence shows that substantial quantities of primary production are not directly assimilated by a trophic process (see, for example, Fransz and Gieskes, 1984). In addition there is some question whether particular phytoplankton species have the same nutritive quality in the context of the grazer-phytoplankton arrangement. Further, recent work has distinguished the tremendous variability in behavior, physiology, and function of phytoplankton, and a growing body of evidence suggests that considerable differences exist in the nutritive quality of certain species or groups of phytoplankton for herbivores (see, for example, Paffenhoffer, 1976). Thus the impression that primary production may in certain instances be underestimated needs to be coupled with an impression that its direct utilization may be overestimated, although the extent of these biases might be accentuated by confounding estimation difficulties in so-called oligotrophic systems with those of systems characterized by high-standing crops of phytoplankton.

The questions associated with primary production are rivaled in number by those associated with higher-level production. Take fish production, which occurs at both planktonic and nektonic stages. At the planktonic stage questions arise regarding the underestimation of growth efficiency and ichthyoplanktonic biomass. Some researchers implicitly assume that the bulk of ichthyoplankton are related to commercial species or that the quality and quantity of ichthyoplanktonic biomass varies with periodicities that are considerably greater than sampling frequencies, thus missing major contributions to its variance. At the nektonic stage questions on production relate to species composition and to underestimation of the growth of reproductive biomass (see Chapter 6). In addition, there is an important potential for densely schooled fish to contribute, under some circumstances, to nutrient patchiness through periodicities in their excretion rates. While energetics of adult-fish growth and mortality are evidently significant, those of the larvae seem to be insignificant, and yet the major fluctuations in fish stocks seem to result from variations in larval abundance.

In sum, an examination of the ecosystem-model approach reveals that a suitable foundation for linking ecosystem models with recruitment variability, not only of fishes but of other organisms, is not yet available. It is not clear at present why this is so, except that the problem seems to involve conceptual insights into ecosystem structure, particularly density-dependentlike and amplifier effects. Including these effects in ocean productivity models and enlarging the scope of the model to take into account major known features that are often not considered, such as the partitioning of biomass into its somatic and reproductive components and the appropriate scope of the model, will obviously require a clever incorporation of the populations dynamics of the involved organisms into ocean productivity studies. This effort may temporarily divert attention from studying flux or bulk transfer of components in the system to an attempt to achieve an understanding of particulate dynamics, which may be integrated at a later point to form an enhanced theory of bulk-transfer system energetics.

Population Dynamics

The study of the population dynamics of oceanic biota is distinct from, but closely parallel to, the study of productivity, as population dynamics takes into account the fate of individual organisms, while productivity takes into account the production and flux of biomass.

Population dynamics may hold important insights for the study of the transfer of biological energy in the sea. It is a subject whose fundamental concerns are the mechanisms associated with the increase, stability, or decrease of animal populations (the same calculus may be used to study any particulate matter). These dynamic changes may be observed in terms of the relation of the number of parents to the number of progeny. Oversimplistically, parents produce progeny that are eventually transformed into parents, and when the second set of parents is equal in number to the first set, the population is stable. When the second set is smaller than the first set, the population declines; when it is greater, the population increases.

The trajectory toward numerical equivalence or difference between the first and second sets of parents lies at the heart of the population-dynamics process in all animals. Yet, between low-fecundity animals (such as mammals and birds) and high-fecundity animals (such as many fish and invertebrates), the nature of the trajectory differs substantially. The difference between high- and low-fecundity trajectories may be seen by considering the sequence of dynamic events in a semelparous population. (The assertions are made for a semelparous population to avoid notational complexity; they could be made for an iteroparous population as well.)

Consider, for a semelparous species, the sequence

$$P_1 \longrightarrow F_1 \equiv Y_1; \quad Y_1 \longrightarrow P_2; \quad P_2 \longrightarrow F_2 \equiv Y_2;$$
$$Y_2 \longrightarrow P_3; \quad P_3 \longrightarrow F_3 \equiv Y_3, \tag{4.1}$$

where P_i is the number of parents, F_i is the fecundity of the parents, and Y_i is the number of young (F_i is identically equal to Y_i at time of birth or spawning). The sequence contains two kinds of transitions. The first is of the form $P_i \rightarrow F_i \equiv Y_i$ and is the egg-production or *fecundity transition*. The second is of the form $Y_1 \rightarrow P_2$ and is the *juvenile-mortality transition*, or the transformation of "just-born" individuals into mature adults of the next generation. Now for low-fecundity species,

$$P_i \cong F_i \equiv Y_i, \tag{4.2}$$

$$Y_i \cong P_{i+1}, \tag{4.3}$$

where $\cong$ means "the same magnitude." The condition of stability for the low fecundity population is

$$P_i = P_{i+1} \tag{4.4}$$

where the equal sign in this case means "statistically equal."

In contrast, for the high-fecundity population we have

$$P_i \ll F_i \equiv Y_i, \tag{4.5}$$

$$Y_i \gg P_{i+1}, \tag{4.6}$$

whereas the condition for stability is identical with (4.4), namely,

$$P_i = P_{i+1}. \tag{4.7}$$

Hence, while the stability conditions for high- and low-fecundity populations are identical, the trajectory between P_i and P_{i+1} must be quite different. In other words, both high-fecundity and low-fecundity cohorts must pass through an identical stability gate (Figure 4.6). The arrival at this gate is tempered by very different sorts of events in high- and low-fecundity populations. High-fecundity populations have much greater potential for stochastic variation than do low-fecundity populations. Yet the actual variability in the former does not appear to be considerably greater than that of the latter. Population control in high-fecundity populations, therefore, must be more finely tuned than that in low-fecundity populations.

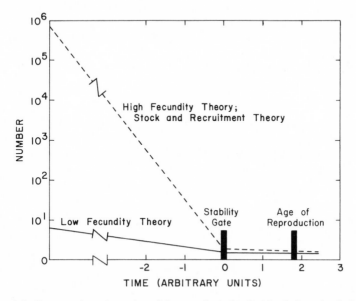

Figure 4.6 Contrast in trajectories of the number of individuals from high-fecundity and low-fecundity populations toward a mean replacement abundance.

Because the trajectory toward relative stability is quite different in the two populations and the mechanisms that govern the approach are also evidently different, one can then wonder at the applicability of a single population-dynamics model for all populations (such as the logistic model) as well as the general applicability of evolutionary interpretations that might derive from the model.

Many ecological population-dynamics models do not make the distinction between low-fecundity and high-fecundity population dynamics. Such models (for example, the Verhulst logistic) are conformable to low-fecundity organisms but not to high-fecundity ones. That is to say, low-fecundity dynamics consider relations of the form $P_i \rightarrow P_{i+1}$, but high-fecundity dynamics, owing to the site of variability, must take account of the *modus ponens*, $P_i \rightarrow F_i \rightarrow Y_i \rightarrow P_{i+1}$, where in particular $Y_i \rightarrow P_{i+1}$ is the recruitment-stock relation as it has been conventionally used.

Difficulty in applying low-fecundity dynamics to high-fecundity animals has been recognized in the fisheries literature for more than seventy years. For example, Hjort (1914: 202) observed that "the opinion generally prevalent hitherto was that the renewal of the stock of fish took place, as in the increase of any human population, by means of a more or less constant annual increment in the form of new individuals; the results here arrived at, however, indicate that this renewal, in the case of the species investigated, is of a highly irregular nature." In one of his most important contributions, which is generally overlooked, Hjort further considered the inadequacy of low-fecundity dynamics in representing high-fecundity populations in more detail (1938).

Ricker (1954: 562) was also concerned with the problem. He pointed out that the Verhulst logistic equation used as a point of departure by earlier population dynamicists was inadequate because it did not take into account the distinction between the dynamics of early life history and that of adult life history. Ricker observed that Verhulst's equation,

$$\frac{dN}{dt} = bK^{-1}N(K - N), \tag{4.8}$$

implied a continual adjustment toward K. This adjustment would have to operate at all ages. He noted that such a compensatory mortality is unlikely in adult fish. (Chapter 6, however, addresses the subject of adult compensation.)

Thus it is evident that the body of population-dynamics theory falls short of addressing the dynamics of high-fecundity species. On one hand, low-fecundity dynamics avoids the details of compensation at the larval and juvenile stages; but on the other hand, recruitment-stock theory avoids the details of compensation during the adult stage. Advances in population dynamics will require reconciliation of the two approaches.

Linkage between Fishery Management and Fundamental Knowledge

Traditional models are important but have limited utility. Their importance derives from their ability to relate average recruitment, yield per recruit, or surplus production to size-specific fishing mortality or stock, given fixed "conditions" or an environment that is stationary over the several-decade interval for which estimation data are available. Limitations of traditional models derive from their inability to accommodate or predict changes in the conditions. Advances in management will require new theory that focuses on and explains the conditions. This theory will be useful for responding to management requirements associated with ecosystem and multispecies management, stock-rebuilding strategies, habitat management, and aquaculture.

To develop new theory, it is important first to consider extant disciplinary knowledge. Our present understanding of ocean productivity needs to be greatly enhanced to be useful for the study of fish-population dynamics. At the same time, our understanding of population dynamics can make a substantial contribution to our understanding of ocean productivity. Moreover, the field of population dynamics needs to address high-fecundity populations if it is to be aligned with ocean-productivity studies in general and fish-population dynamics in particular, as it appears to be oriented more to low-fecundity populations, which involve a different set of problems.

Association of the theory with resource management requires that its evolution take into account the series of decisions that lie at the heart of any management process. Because the technical aspects of management can be represented as a series of decisions, it would appear that the foundations for organizing the applied and fundamental problems may be better appreciated in a decision-theoretic setting. In considering such a setting, the question arises as to whether a series of management decisions is "optimal" or "suboptimal," and if it is suboptimal, how can it be improved? The evaluation of optimality needs to take into account the structure of individual decisions. Each decision comprises (1) a pool of information relevant to the decision, (2) a set of rules that converts the information into an intended action, and (3) implementation of the action.

In simple settings all concerned agree on the nature of the information pool, the rules, and the strategies or tactics for implementation; hence decision making is a fairly straightforward process. Decision making in resource-management settings, however, is typically not straightforward, because the decisions must be reached by balancing the very different interests and backgrounds of groups of individuals that generally participate in the decision-making process. The different interests and backgrounds tend to generate disagreement on the quality of the information, on the rules, or on the implementation.

Situations in which there are considerable differences in viewpoints regarding the information base, the rules, and the implementation usually result in sets of decisions that are less than globally optimal. The issue, then, is to what extent less-than-optimal decisions result from a lack of information or from differences that would exist in perfect-information situations. In other words, could the decision process move more toward the global optimum, however it may be defined, if information were increased?

This question is obviously nontrivial. The framework for its answer rests in the application of decision theory, which can identify in the abstract not only the kinds of information required by management but its benefits and costs as well. Specific rudimentary applications of decision theory to fishery management were discussed by Rothschild and Heimbuch (1983). They described the decision problem as being comprised of three components: (a) nature's strategy, (b) the decisionmaker's strategy, and (c) the consequences or payoffs.

Nature's strategy consists of chance events of the form

$$\theta = (\theta_1 = i, \theta_2 = i, \ldots, \theta_m = i), \qquad i = 1, 2, 3,$$

where, for example, $i = 1$ refers to a good year class, $i = 2$ to a poor year class, and $i = 3$ to a catastrophically poor year class. Thus in eight years of observations for a particular stock, θ might be represented by (1, 1, 2, 2, 1, 2, 1, 3).

The second component, the decisionmaker's strategy, is denoted by a sequence of actions or decisions of the form

$$A = (A_1, A_2, \ldots A_n),$$

where each A_i generally refers to an action or a decision. A most typical form of the decision would be to set some value of $\mathbf{F}_i$, the age or size-specific vector of fishing mortality in the ith year. For example, fishing mortality could be high (H), low (L), or zero (Z).

The third component, the payoff, or the value of the decision, generally depends on a particular state of nature θ_i and action A_i. Intuitively, the payoff should be quite low if the state of nature is such that recruitment is catastrophically poor and the resulting action is a very high F_i.

These elements are usually integrated in the form of a decision tree. A hypothetical tree in Figure 4.7 shows, for example, that the states of nature could be either good recruitment, poor recruitment, or catastrophic recruitment. The decisionmaker's task is to guess or otherwise estimate by experimentation or a sampling process the state of nature and the appropriate action, which in this case might be implementation of high fishery mortality, low fishery mortality, or zero fishery mortality.

Because the state of nature is a chance event and perfect information is not available, the manager does not always choose the correct strategy. The payoffs or costs of good and poor choices are also shown in Figure 4.7.

The simplest approaches to deciding on the appropriate decision would involve maximin management, expected-monetary-value (EMV) management, and expected-utility management. The maximin approach is least demanding of information and is most conservative. It chooses the decision strategy that minimizes the maximum loss. A maximin ap-

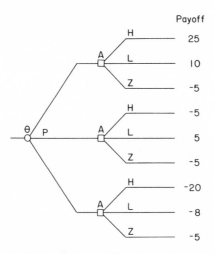

Figure 4.7 Hypothetical decision tree. The figure shows nature's strategy of choosing a good (*G*), poor (*P*), or catastrophic (*C*) year class; the decisionmaker's action *A*, of implementing high (*H*), low (*L*), or zero (*Z*) fishing mortality; and the payoff for any action (if, for example, nature chooses *P* and the decisionmaker responded with *L*, then the payoff would be 5 units). (From Rothschild and Heimbuch, 1983: fig. 2.)

proach would involve choosing Z, because -5 units would be the maximum lost independent of θ. But if some information existed on θ — for example, if it was known that

$$P[\theta = G] = 0.45, \qquad P[\theta = P] = 0.45, \qquad P[\theta = C] = 0.10$$

— then the expected or average value of the variance, given the actions H, L, or Z, would be

$$E[V|A = H] = 7.0, \qquad E[V|A = L] = 5.95, \qquad E[V|A = Z] = -5.0.$$

Hence the additional information would result in a different strategy, H. And H would result, on average in a higher payoff than Z. In fact, the differences would be twelve monetary units.

But is the choice of H really that much better than the choice of Z? Computing variances gives

$$\text{Var}[V|A = H] = 283.5, \qquad \text{Var}[V|A = L] = 27.2,$$
$$\text{Var}[V|A = Z] = 0.$$

Now the choice depends on one's attitude toward risk. Is it preferable to have a high payoff that is quite variable or a small loss that is certain?

This elementary example is instructive. A real-world decision system would involve decision trees of much greater complexity. The adornment of such trees with even reasonable guesses of the states of nature, the decisionmaker's strategies, and the payoffs would be unfeasible in many cases because most fishery management is not considered in a decision-theoretic context. The idea, then, is that the decision-analysis framework, at least at present, constitutes a framework that defines the kinds of technical (as opposed to political, for example) information required for fishery management. The framework shapes questions that depart from those asked by the traditional models. Such questions relate, among other things, to the *probability* of various events, like good and poor year classes, and to the *probability* of detecting or predicting good or poor year classes, as well as to the *costs* of information acquisition. Whereas in the past we might have been satisfied to predict some average value of recruitment, we are now more concerned with understanding the probability of recruitment, a much more demanding task.

The decision-theoretic approach to fish-stock management is attractive from both the applied and the fundamental points of view. The decision-theoretic model sets the stage for developing the kinds of practical questions and answers required by resource managers, whether they are managing a particular fishery or considering constructing a power plant that might affect the recruitment of a particular fishery. The same questions identified in the applied decision-theoretic model are basic for building a general theory explaining the mechanisms that cause populations to vary. Thus the development of fishery-management information is an applied problem, but the required information is most fundamental in nature.

An Ecosystem Approach to the Recruitment-Stock Problem

To develop some notions that relate to estimating the probability of good or poor year classes or the conditions associated with them, it is helpful to rely on yet another organizing structure. Such a structure needs to take into account axiomatically (1) the complexity of the system, (2) the fact that it includes many species, (3) the fact that many of the species have life-history stages that vary as much between species as within them, and (4) the fact that variability may arise as much from environmental variables as from biotic variables.

A system that takes into account these properties is shown in Figure 4.8. Note that the ith species $(i = 1, \ldots, N)$ each have k_j,

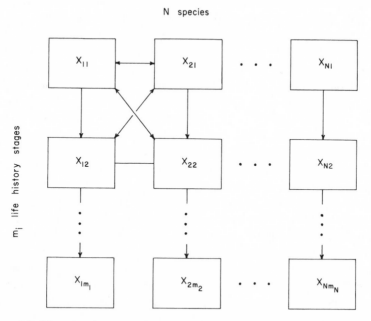

Figure 4.8 The complex ecosystem consisting of N species each with m_i life-history stages set in the physical environment.

Species

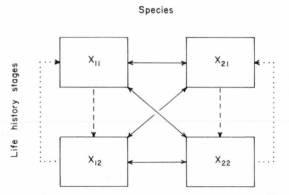

Figure 4.9 The simplest nontrivial ecosystem consisting of two species with two life-history stages. The *dotted arrow* refers to egg production; the *dashed arrow,* to recruitment. The *solid arrows* refer to possible interspecific interactions. Note that within any species it is necessary to consider variability in terms of both egg production and recruitment as well as in terms of interspecific interaction.

$j = 1, \ldots, m_i$ life-history stages. If all the species have the same number of life-history stages, then the possible number of interactions among the species' life-history entities would be $NK(NK - 1)$. Thus if each entity and all its interactions with other entities were to be studied, as well as, say, three environmental variables, there would need to be $3(NK + NK(NK - 1))$ possible subjects of study. Although some of the interactions could be rejected, it is unlikely that the total number would differ from a very large number. Suppose there were five species, each with two life-history stages; then there would be $3(10 + 90) = 300$ possible entities for study.

The analysis is telling. Obviously many aspects could warrant study. It is difficult to say what is important and what is not, even in a simple system — particularly if the importance of the variables is changing. Further, although it may be possible to understand or predict the behavior of a single organism, the increase in the variables of interest is at least as great as the square of the number of organisms with reasonably complex life histories. Thus it is unlikely that resources could be devoted to the study of the system as a whole, and so only a small subset can be studied. But *which* subset should be studied, and *why* is one particular subset better than any other? The converse of this problem relates to our capacity to understand the system. What are the bounds for understanding such a complex system, and what are our targets for predictive performance?

These questions suggest the need for a set of rules, a syntax, to collapse the problem to an intermediate level of complexity. What these rules might be is not at present clear, but some insights might be gained by examining the converse of the most complex ecosystem depicted in Figure 4.8, the simplest nontrivial ecosystem. This ecosystem (Figure 4.9) has two species, each with two life-history stages. Thus there are four compartments and twelve possible interconnections. Again, consideration of the environmental variables suggests $(4 + 3 \times 4)\,3$, or 48, entities. But even the full consideration of this simple system would be quite difficult, as the information on interspecific interaction is minimal.

The problem is that even in this simple configuration it is hard to appreciate the workings of the system. We would have to examine what we already know to determine how this knowledge can be extended, which places us back at the centerpiece of fish-population-dynamics theory, the recruitment-stock model. A more detailed examination of the recruitment-stock concept will reveal the mechanisms that both stabilize a population and contribute to its stochastic behavior. Clearly such an examination will improve the interpretation of the traditional models and provide insights into the formulation of new models.

5

The Recruitment-Stock Paradox

The famous recruitment-stock relationship of fishery biology* is an important component of the study of high-fecundity population dynamics. The theory linking recruitment and stock predicts the number of recruits that will be produced by the population for any given stock size, taking explicit account of neither associated species nor the physical environment. The theory has two classic branches, both of which predict that at relatively low stock size, recruitment increases nearly in proportion to stock size. One branch, developed by Ricker (1954), suggests that at relatively high stock sizes recruitment will decline; the other branch, developed by Beverton and Holt (1957), suggests that at relatively high stock sizes recruitment does not decline but rather approaches an asymptotic value (Figure 5.1).

The curvilinear nature of the classic Ricker and Beverton-and-Holt recruitment-stock curves shown in Figure 5.1 is evidently an important property of the population-dynamics process; reproductive efficiency (the number of recruits produced per spawning fish) is enhanced when the stock declines at a low level of abundance, and it diminishes when the stock increases to a high level of abundance, conferring to the population a certain stability. In other words, the curvilinearity of the recruitment-stock function ensures a certain resilience in the stock.

The degree of nearly proportional increase in recruitment at relatively low stock levels is thought to reflect a *density-independent* population response, whereas the decline in the rate of recruitment increase at intermediate or high population levels is thought to reflect a *density-dependent*

---

* The literature might be entered through Ricker (1954, 1958) and Beverton and Holt (1957). *Fish Stocks and Recruitment*, edited by Basil Parrish (1973), contains a number of important papers. Among these Paulik (1973) provides a good overview. Cushing (1981) places the problem in biological perspective.

Figure 5.1 Examples of classic recruitment-stock curves. Curve (*a*) is a Ricker curve; curve (*b*) is a Beverton-and-Holt curve. Ricker (1973) gives a detailed comparison of these curves.

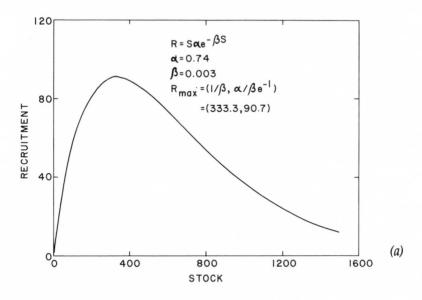

(*a*)

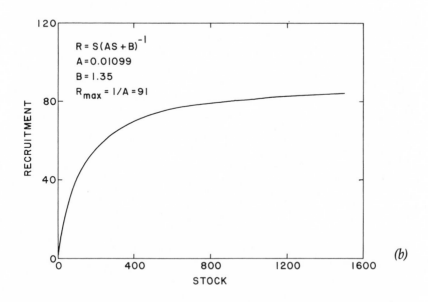

(*b*)

response. That is, at intermediate or higher stock densities, reproductive effectiveness decreases because of density-dependent population responses.

Recruitment-stock theory is of great potential utility for the resource manager or the resource-policy decisionmaker. Any regulation that affects the catch must in general affect the fishing-mortality vector (see Rothschild, 1977), which in theory affects the biomass curve, which in turn affects the spawning biomass, which determines the magnitude of recruitment.

Armed with a recruitment-stock theory, the resource manager can determine the recruitment effect of any management strategy. If such a theory did not exist, the manager would lack sufficient information to control abundance, except in the short run. In that case the manager's actions would be dominated by anticipated recruitment variability, which can be of unexpected magnitude, even if a stabilizing population-dynamics process exists.

Despite the apparent importance of the recruitment-stock relationship in population control, a striking relation between the magnitude of recruitment and spawning stock is not reflected in the body of empirical data (Figure 5.2 shows some examples from Garrod, 1982). This situation is a paradox, because it seems that a mechanism critical to population stabilization would reflect a better correspondence between data and theory, even taking account of measurement error. How can there be so little evidence for a relation of recruitment to parent stock when the very existence of such a relationship is so critical to population stability?

Investigation of this paradox requires an appreciation of the complexities of recruitment-stock theory as well as its subtleties, which can be gained only by considering the theory in its most elementary form. Figure 5.3 shows the relationship between recruitment and stock for a hypothetical stock "recruited" at age zero and attaining maturity at age two. As can be seen, the relationship is an average of a complex series of events. For example, the 1962 recruitment-stock relation nominally involves only the spawning stock in 1962 and its resulting recruitment; however, the spawning stock in 1962 comprises descendants from recruitment in 1958, 1959, and 1960. Furthermore, it takes two to four years for recruits from 1958, 1959, and 1960 to "reach" the 1962 spawning aggregation. The process by which the recruits reach the spawning aggregation — that is, the relation of recruits to spawning fish — may be as important as the relation of stock to recruitment, but it has not received serious treatment in the literature. Note that Figure 5.3 is implicitly a steady-state model. The passage of a large year class through the system portrayed in the figure would raise additional questions: for example, if the 1960 year

Figure 5.2 (a–g) Recruitment-stock observation from several stocks. The curves are the recruitment-stock function suggested by Shepherd (1982), which minimizes the impression of a "dome" in the relationship. (Redrawn from Garrod, 1982: fig. 5.)

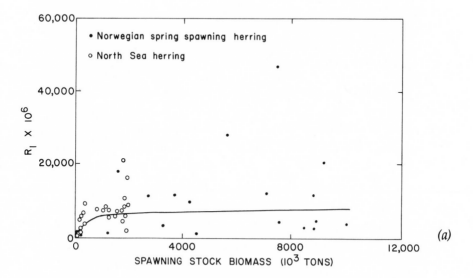

(a)

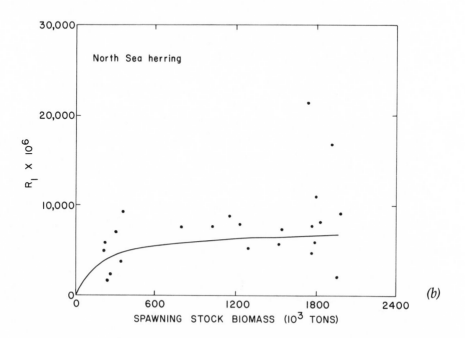

(b)

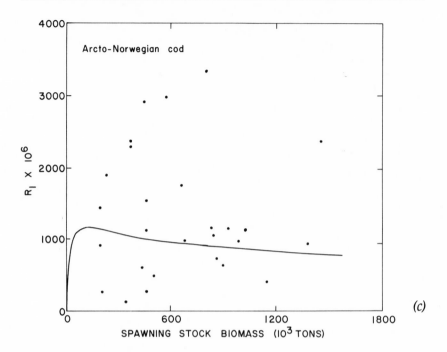

(c)

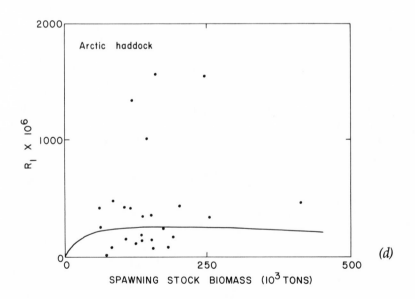

(d)

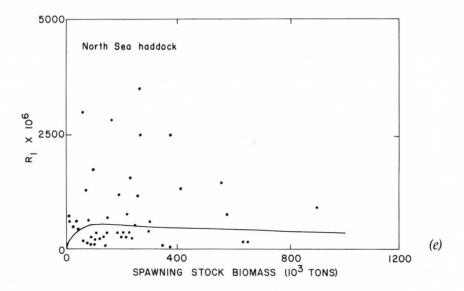

(e)

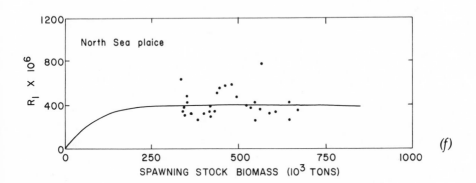

(f)

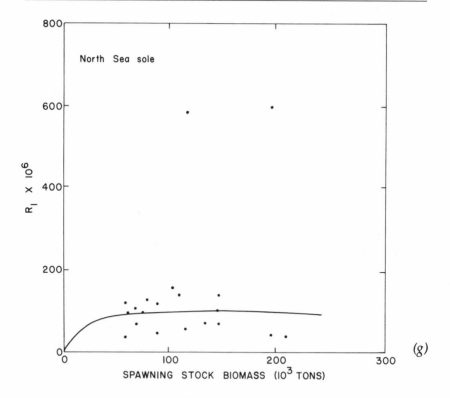

(g)

class were exceptionally large, how would it affect adjacent year classes? Finally, the system is embedded in a plexus of biotic and abiotic variables; it can be imagined that any biotic or abiotic perturbation occurring at any one of the X's, or among any group of X's, could be amplified as it moved through the plexus, creating considerable interpretive difficulty with respect to the site of its causation.

It is easy to see that the relation between recruitment and stock in a particular year involves events not only in that year but in previous years, and that these events are linked in at least three ways: (1) by the relation of stock to recruitment (the juvenile-mortality transition); (2) by the relation of recruitment to stock (the fecundity transition); and (3) by the relation of stock and recruitment to the physical and biological environment. Although many aspects of these three relationships have been considered in the thirty years since Ricker proposed his theory, the literature has tended to concentrate on the relation of recruitment to stock in a setting independent of interannual relationships.

The general nature of the recruitment-stock relation was well articulated by Paulik (1973). He observed that the relation was merely that of

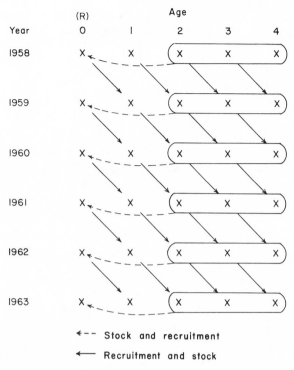

Figure 5.3 Recruitment-stock relation in a population of fish where all individuals are recruited at age zero, become mature in their second year, and die at the end of their fourth year. The X's represent the number of fish alive in each age group in each year; the *circled* X's, the spawning stock in each year; the *dashed arrows*, recruitment-stock relation in each year; and the *solid arrows*, the movement of each year class through the population. Note that recruitment in 1962 depends on the recruitment-stock relation in 1962, which is evaluated at the 1962 spawning-stock-size level. The latter is a function of, among other things, the magnitude of recruitment in 1958, 1959, and 1960.

the number of individuals in a stock at the time of spawning (actually, the number of eggs spawned by the parent stock) to the number of individuals in that stock at some later date, when the individuals reached recruitment age. In other words, in Paulik's approach we are concerned with the number of individuals in each cohort, as it proceeds through its life, not the labels (for example, "eggs," "adults") applied to the individuals. Taking account of Neave's (1952) notion of compensation, Paulik linked the number of spawning individuals $N_{t+\tau}$ to the number of eggs N_t by

$$N_{t+\tau} = N_t g_t(N_t, e_t), \tag{5.1}$$

where τ is the age of recruitment and g is a growth function depending on population size N_t and environmental conditions e_t.

Using these definitions and setting $N_{t+\tau} = R_{t+\tau}$ and $N_t = S_t$, Paulik obtained the general recruitment-stock relation

$$R_{t+\tau} = S_t g_t(S_t, e_t). \tag{5.2}$$

If $E(e_t) = 0$, the average value for recruitment is

$$R = Sg(S). \tag{5.3}$$

Equation (5.3) is the general form of the recruitment-stock relation, which accounts for compensation by setting g to be a monotonically decreasing function of S. Paulik observed that (5.3) represents a family of extinction curves — (a) the inverse square-root law, (b) the inverse two-thirds law, (c) the rectangular hyperbola, (d) the exponential curve, (e) the parabola, and (f) the straight line — that was discussed by Turner, Monroe, and Lucas (1961). Paulik called attention to the fact that the most commonly used recruitment-stock curves were among these extinction curves. Hence the Ricker curve is exponential, with

$$g(S) = \alpha e^{-\beta S}, \tag{5.4}$$

and the Beverton-and-Holt curve is a rectangular hyperbola, with

$$g(S) = (AS + B)^{-1}. \tag{5.5}$$

The discussion thus far has demonstrated that in a very general way the classic recruitment-stock relationships are but special cases of a large family of mathematical functions. The following elementary exposition of these special cases elucidates the extent to which recruitment-stock theory is understood, as well as the avenues that might be fruitful for its further consideration.

Density Dependence

The notion of compensation mentioned above can be explained in terms of so-called density dependence. The idea is that the lower left-hand corner of a recruitment-stock graph shows that recruitment is nearly proportional to stock; hence their relationship is virtually linear in this "density-independent" portion of the graph. In the Ricker curve the slope of the density-independent portion is the constant α. In the Beverton-and-Holt curve, it is B^{-1}. As the stock increases in size, the rate of recruitment begins to decline. This decline in recruitment is believed to reflect the operation of density-dependent factors in the relation of stock to recruitment. In other words, as stock increases in size, the "efficiency" of recruitment is reduced.

Cushing (1971) thought that clues to the particular form of density dependence might be developed from

$$R = kS^b, \tag{5.6}$$

where R is the magnitude of recruitment, S is the magnitude of stock, and b and k are constants. In particular, he thought that the magnitude of constant b, which reflects the degree of curvature in the recruitment-stock relationship, represents the degree of density dependence in the recruitment-stock relation. Cushing obtained average values for b in several stocks and compared these with fecundity; he showed that as fecundity increased, the index of density dependence tended to decline. Cushing and Harris (1973), using analogous reasoning, estimated the value of the exponent in (5.4), which they reparameterized as $B\overline{P}$. They obtained estimates of $B\overline{P}$ for a variety of stocks, which ranged from 0.45 in Atlantic herring to 1.75 in gadoids (Figure 5.4). If $B\overline{P}$ is an index of

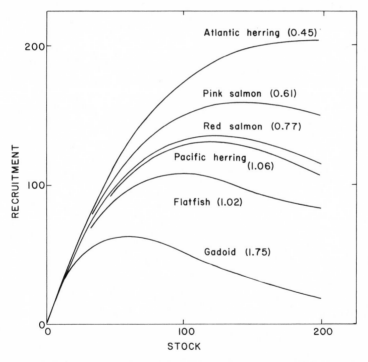

Figure 5.4 The form of recruitment-stock curves by groups of fish, with constants averaged. The values of $B\overline{P}$ are included in parentheses. Note that low values of $B\overline{P}$ are nearly asymptotic, whereas high values of $B\overline{P}$ result in a dome-shaped curve. (From Cushing and Harris, 1973: fig. 114. Reprinted with permission of ICES.)

resilience capability or compensation, the gadoids are much more resilient than the herring in that, at large stock sizes, declining herring stocks would result in declining recruitment, but declining cod stocks would result in increased recruitment.

To study the compensatory behavior of recruitment-stock functions in more detail, consider the traditional mortality model, which gives (as will be recalled from Equation 3.2) the number of individuals alive in a cohort N_t at any time t, given an initial number of individuals N_0 and a coefficient of mortality M:

$$N_t = N_0 e^{-Mt}. \tag{5.7}$$

To see the connection between this ubiquitous mortality-rate model and recruitment-stock theory, we revise the notation as follows:

a. Let $N_t = R$, or the number of recruits.

b. Let $N_0 = e \times s = S$, where e is the number of eggs per spawning fish, s the number of spawning fish, and S the number of eggs produced by the population. (*Note:* S and N_0 will be used interchangeably.)

c. Let $Mt = \mu$, where t is the number of years between hatching and recruitment and is scaled to be equivalent to unity.

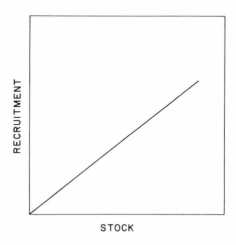

STOCK

Figure 5.5 A density-independent recruitment-stock relation, $R = Se^{-\mu}$. The slope of the recruitment-stock function is obviously $e^{-\mu}$. Although simple, the graph demonstrates and emphasizes the nature of density independence.

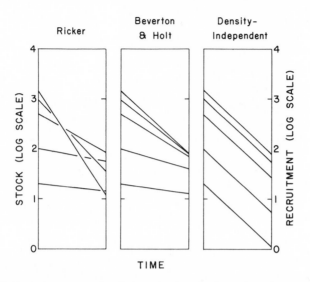

Figure 5.6 Recruitment-stock relations in the number-time plane showing the compensatory behavior of various functions. The nature of compensation in each theory is shown by the behavior of the family of linearized trajectories associated with the theory. Compensation is greatest in the Ricker function, in that the linearized trajectories actually cross. There is no change in the rate of decline as a function of the initial numbers in the density-independent form. The Beverton-and-Holt form approaches a constant asymptotic value.

The notational revision yields

$$R = Se^{-\mu}, \tag{5.8}$$

which is in fact a density-independent recruitment-stock relation (Figure 5.5). The fact that (5.8) is density-independent is demonstrated by the constancy of recruitment efficiency; it does not change with stock size. Recruitment is always a fixed proportion of stock size, a situation unlikely to exist in nature.

It is possible to compare recruitment-stock relationships with regard to their density-dependent or compensatory properties. This comparison is more striking in the number-time plane than in the conventional number-number plane. Figure 5.6 shows recruitment-stock relations in the number-time plane. The principle difference among these relations is in the way that mortality rates change as a function of the initial density of animals. In the density-independent curve, mortality rates do not change; they are constant. Hence the density-independent curve in itself does not reflect compensation. In the Beverton-and-Holt curve, mortality rates continue to increase as population increases. The same is true in the Ricker curve, except that in it the rates increase faster.

Before we carry this analysis too far, it is important to recognize that in Figure 5.6 the number-time plane contains all the information in the number-number plane. It reflects, however, the cost associated with making time implicit in the traditional recruitment-stock relationship, in the sense that there are alternative ways of joining any two fixed points, and study of these alternatives might lend considerable insight into the mechanisms affecting the relation of recruitment to stock.

Young-Fish Mortality Functions

The literature on recruitment and stock does not sufficiently emphasize the simple notion that recruitment-stock theory is no different than a theory that attempts to account for the mortality of fish between spawning and recruitment time. Authorities have had differing viewpoints on this relationship; here we explore the classical point of view, derivations from the classical point of view, and approaches toward the study of intermediate events.

Classical Theory

Harris (1975) provided an integrated view of the classical theory by building his derivations in the spirit of Equations (5.3) and (5.7); in fact, he used both (5.7) and its derivative $dN/dt = -MN$. He observed that (5.7) is a density-independent function. To reflect density-dependent behavior, M would have to be a function of population density. That is, M should increase as the number of individuals in the population increases. Harris suggested that the Beverton-and-Holt assumption,

$$\mathfrak{M} = M_1 + M_2 N, \tag{5.9}$$

was the simplest assumption representing density-dependent mortality. (An even simpler assumption would be to let $\mathfrak{M} \equiv MN$, which would result in the differential equation $dN/dt = -MN^2$.) Equation (5.9) implies that the mortality rate $\mathfrak{M}$ increases linearly with N. The intercept is M_1 and the slope is M_2. Insertion of (5.9) in $dN/dt = -\mathfrak{M}N$ yields the differential equation for the Beverton-and-Holt recruitment-stock equation, which will be recognized as a Bernoulli differential equation:

$$\frac{dN}{dt} = -(M_1 + M_2 N)N. \tag{5.10}$$

Integrating (5.10) yields (5.3) with (5.5) as $g(S)$. A more explicit definition of the constants is sometimes useful. Following Harris, we have

$$R = \left\{ \left[\frac{M_2}{M_1} \right] [e^{M_1 \tau} - 1] + \left[\frac{e^{M_1 \tau}}{es} \right] \right\}^{-1}, \tag{5.11}$$

where $es = N_0$ refers to the number of eggs and τ refers to the recruitment age.

To consider the alternative classical model, the Ricker model, Harris expressed the Ricker assumption as

$$M = M_0 + k N_0 \tag{5.12}$$

and substituted this value into the basic mortality equation. Thus using $es = N_0$,

$$\begin{aligned}
R &= N_0 e^{-M\tau} \\
&= N_0 e^{-(M_0 + k N_0)\tau} \\
&= N_0 e^{-M_0 \tau} e^{-k N_0 \tau} \\
&= \{e e^{-M_0 \tau}\}\{s\}\{e^{-kes\tau}\}; \tag{5.13}
\end{aligned}$$

letting $\alpha = e e^{-M_0 \tau}$; $\beta = ke\tau$, we have

$$R = \alpha s e^{-\beta s},$$

which is (5.3), with (5.4) as $g(S)$.

It is important to use care in interpreting (5.13), as it depends only on the *initial* number of fish, a function of N_0, and not on the number of fish at any time. This was indeed the interpretation envisaged by Ricker (1954: 610), who viewed the close relation between N_0 and M as resulting from cannibalism:

> Of all the methods of population regulation listed earlier, cannibalism is the one in which abundance of the control agent is most closely and inseparably allied to that of the population controlled. That is, an increase in the mature stock not only increases the number of eggs laid of young born in a given reproductive season, but it also decreases the rate of survival of those young.

Harris, however, did not accept cannibalism of very young fish as a likely possibility, even noting that some recruitment-stock curves, such as that for arcto-Norwegian cod (for example, Garrod, 1977), were dome shaped. He explored differential equations of the form $dN/dt = g(N)N$ (cf. Equation 5.3) and concluded that as stock increased, recruitment either increased or approached an asymptote. (It is now known that cannibalistic predation is quite important in a number of stocks.)

Derivations from Classical Theory

Many variations have developed on the theme of $R = Sg(S)$ since Ricker and Beverton and Holt. (In fact, Ware, 1980, and Ulltang, 1984, combined the two approaches.) Some examples are given below (interesting further examples may be found in Csirke, 1980; MacCall, 1980a; Garcia, 1983; and Tang, 1985).

Beverton and Holt on Ricker. Beverton and Holt noticed that the Ricker curve could be written as a composite of two age-specific mortality rates. They observed that the prerecruitment life of the cohort could be partitioned into an early stanza of relatively high mortality, μ_1, and a late stanza of relatively low mortality, μ_2. If the time interval of the early stanza is $t_c - t_0$, where t_c is the transition point between the two stanzas and t_0 is the time of spawning when age is equal to zero, and if the time interval of the second stanza is $t^* - t_c$, where t^* is the recruitment age, then

$$R = N_0[e^{-\mu_1 t_c}][e^{-\mu_2(t^* - t_c)}]. \tag{5.14}$$

Note that (5.14) does *not* reflect density-dependent mortality. Beverton and Holt introduced density dependence by assuming

$$t_c \propto \frac{1}{W_\infty} \propto \frac{1}{\text{food eaten}} \propto \frac{\text{density}}{\text{of}} \propto N_0. \tag{5.15}$$
$$\text{fish}$$

In other words, as N_0 increases, the density of young fish increases, thereby reducing the food eaten per young fish, which decreases their ultimate size or growth rate, which in turn increases t_c, or the exposure to high mortality rate.

Thus by assumption (5.15), the additional mortality $(\mu_1 - \mu_2)$ and its intensity $(\mu_1 - \mu_2)t_c$ are proportional to the number of spawners. Letting β be the constant of proportionality, we have

$$(\mu_1 - \mu_2)t_c \propto N_0, \qquad (\mu_1 - \mu_2)t_c = \beta N_0. \tag{5.16}$$

In addition, let

$$e^{-\mu_2 t^*} = \alpha. \tag{5.17}$$

Now, using (5.14), (5.16), and (5.17), we have

$$
\begin{aligned}
R &= N_0[e^{-\mu_1 t_c - \mu_2 t^* + \mu_2 t_c}]\\
&= N_0[e^{-t_c(\mu_1 - \mu_2)}][e^{-\mu_2 t^*}]\\
&= \alpha N_0 e^{-\beta N_0}, \tag{5.18}
\end{aligned}
$$

which is the Ricker recruitment-stock relationship, (5.13), parameterized in terms of eggs rather than stock. We need to be careful, though, because despite the fact that (5.13) and (5.18) are identical, apart from the constant, they are arrived at by different assumptions on density dependence. To see this, let $t_c - t_0 = T_1$ and $t^* - t_c = T_2$. Now (5.16) and (5.17) can be solved for μ_1 and μ_2, yielding

$$\mu_1 = \frac{1}{T_1} \beta N_0 - \frac{\log \alpha}{T_1 + T_2}, \qquad \mu_2 = -\frac{\log \alpha}{T_1 + T_2}. \tag{5.19}$$

In other words, for fixed parameter values α and β, for each level of N_0, and for a fixed prerecruitment age $T_1 + T_2$, the mortality rates depend on the relationship of T_1 to T_2, which is a more general assumption than (5.12).

Harris (1975) on kth-power density dependence. Several variations on the mortality curves have been developed. For example, Harris considered the generalization of the derivative of (5.7) to raise N to some exponent, namely,

$$\frac{dN}{dt} = -\mathfrak{M} N^k N. \tag{5.20}$$

Noting that the special case $k = 0$ is equivalent to (5.7) and integrating, he obtained

$$\left. \frac{1}{k \mathfrak{M} N^k} \right|_{N_0}^{R} = \tau \tag{5.21}$$

or

$$R = N_0 \frac{1}{[1 + \mathfrak{M} K \tau N_0^k]^{1/k}}, \tag{5.22}$$

the asymptote of (5.22) being

$$R^* = \frac{1}{[\mathfrak{M} k \tau]^{1/k}}. \tag{5.23}$$

Cushing (1975a) used a special case of (5.21) (namely, he set $k = 1$) to consider natural mortality in plaice and to generate implicitly a plaice recruitment-stock curve. Cushing presented his result in the number-time plane, showing the rapid decline in number in the earliest life-history stages.

Power functions. Paulik generalized various forms of the Ricker model. He observed that Ricker's curve could be raised to an exponent,

$$R = \alpha S e^{-\beta s^\delta}, \tag{5.24}$$

which would affect the nature of the right-hand portion of the Ricker curve. He also pointed out that Cushing's curve,

$$R = k S^b, \tag{5.25}$$

could be considered more generally as

$$\frac{R}{S} = \alpha - \beta s^\delta. \tag{5.26}$$

Generalization of the asymptotic curve. Paulik generalized the Beverton-and-Holt curve so that it could be asymptotic to an asymptote of any form (it might, for example, be linear, with a nonzero slope). To obtain the generalization, substitute (5.5) into (5.3) and multiply the numerator and denominator by α:

$$R = S \frac{1}{\alpha S + \beta} \frac{\alpha}{\alpha} = \left[\frac{S\alpha}{\alpha S + \beta} \right] \frac{1}{\alpha} = \left[\frac{1}{1 + \dfrac{\beta}{\alpha S}} \right] \frac{1}{\alpha}. \tag{5.27}$$

Now express the asymptote as $\alpha^{-1} = A$ and the rate of approach to the asymptote as $\alpha/\beta = r$, giving

$$R = \frac{A}{1 + \dfrac{1}{rS}}. \tag{5.28}$$

Paulik cited Hammack (1969) as using two functions in (5.28), the exponential function giving

$$R = \left[\frac{1}{aJ^b} + \frac{\beta}{S} \right]^{-1}, \tag{5.29}$$

and the linear function giving

$$R = \left[\frac{1}{a + bJ} + \frac{\beta}{S} \right]^{-1}. \tag{5.30}$$

The generalized asymptotic regression can then be used to demonstrate trends in environmental variables that might affect the asymptote.

Approach to the Study of Events Intervening between Stock and Recruitment

Of the theories discussed above, none has taken particular account of the events intervening between stock and recruitment. The model of Shepherd and Cushing (1980) moves in this direction because it deals with both the nutrition of the larvae in terms of their growth and that component of larval mortality caused by predation. In addition, because of its orientation to the more classical recruitment-stock theory, the model provides a convenient transition to more event-oriented models.

According to Shepherd and Cushing, events critical to establishing the magnitude of a year class occur when the larvae weigh between W_0 and W_1 g. They also assert that larval mortality results from an interaction between the density of larval food and the intensity of predation. They assemble their argument by expressing the time derivative of growth as

$$\frac{dw}{dt} = \frac{WG^*}{1 + N/K},$$ (5.31)

where G^* is the maximum growth rate when food is abundant, N is population size, and K is a constant relating to the carrying capacity of the system. When N is about equal to K, then the growth coefficient $G(N)$ is approximately equal to $G^*/2$.

The time rate of change in the number of the population is

$$\frac{dN}{dt} = -\mu N.$$ (5.32)

Using the chain rule to eliminate dt from (5.31) and (5.32) and with a partial fraction expansion, Shepherd and Cushing obtained

$$\frac{dW}{W} = -\frac{G^*}{\mu}\left[\frac{1}{N} - \frac{1}{N + K}\right] dN.$$ (5.33)

Integrating the left-hand side through the critical size $W_0 \rightarrow W_1$ and the right-hand side through the number of the population at W_0, N_0 and at W_1, N_1 yields

$$\log\left(\frac{W_1}{W_0}\right) = -\frac{G^*}{\mu} \log\left[\frac{N_1}{N_1 + K}\middle/\frac{N_0}{N_0 + K}\right].$$ (5.34)

Because W_1/W_0 is a predefined constant, as is μ and G^*, a new parameter can be defined:

$$A = \left(\frac{W_1}{W_0}\right)^{-\mu/G^*} = e^{(-\mu/G^* \log W_1/W_0)} = e^{-\mu T_0}, \tag{5.35}$$

where T_0 is the time that larvae would grow from W_0 to W_1 if food were not limiting. Hence by definition A is unrelated to density-dependent effects of feeding.

Using the notation in (5.34), we can rewrite (5.35) as

$$N_1 = \frac{AN_0}{1 + (1-A)N_0/K}, \tag{5.36}$$

which is the number of larvae that survive through the critical period, given the density-independent component of mortality, A, and the density-dependent parameter, K. At very high population levels N_0 will be much greater than K, and hence N_1 depends on both A and k via

$$N_1 \cong \frac{A}{1-A} K, \tag{5.37}$$

whereas at very low population levels N_0 will be much less than K, yielding

$$N_1 \cong AN_0. \tag{5.38}$$

Thus Shepherd and Cushing generated the discussion of the interaction between intensity of predation on larvae and adequacy of larval nutrition. In their model the problem of surviving the critical period is not simply one of predation or larval nutrition but comprises the interaction of the two. The way in which these events interact is shown in Figure 5.7.

Several features of the model bear additional comment; these relate to the fact that predictions that would be derived from the model depend on the assumptions utilized in its formulation. Notably, growth varies as a function of food density, which is measured in terms of the time required for fish weighing W_0 g to attain W_1 g. This is analogous to the assumptions specified in (5.15). Recognize that the assumption is specific to the prey and excludes consideration of the predator. That is to say, mortality is increased when N_0 increases because the exposure to predation is increased, not because the predators become more efficient as prey density increases. Note also that changes in temperature could increase (or decrease) growth, generating a situation that might mimic, at least temporarily, the assumptions in (5.15) even though density of predators, prey, and prey food resources is held constant.

Figure 5.7 The relation of *A* (inversely proportional to predation mortality) to *K* (carrying capacity). Part (*a*) shows the effect of varying *A* with *K* constant; part (*b*) shows the effect of varying *K* with *A* constant. (From Shepherd and Cushing, 1980: figs. 3 and 4. Reprinted with permission of ICES.)

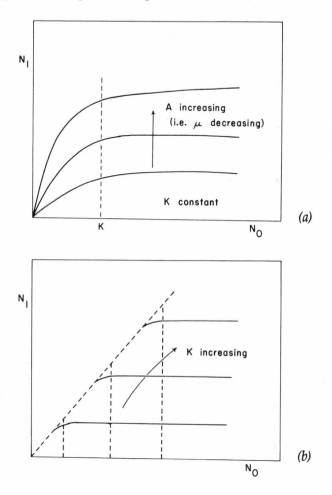

Dynamics

So far, discussion of the recruitment-stock problem has tended to concentrate on its static aspects, as is customary in most recent analyses. Discussed in early analyses, the recruitment-stock system is not static but dynamic. The magnitudes of stock and recruitment are constantly changing. Stock levels in adjacent years are likely to be correlated; hence if the recruitment-stock relation holds, recruitment values in adjacent years should be correlated as well.

One of the first studies of correlations or time-series relationships among "year classes" in fisheries was published by Aage J. C. Jensen in 1927. Jensen studied catch anomalies of the Bornholm herring fishery and reported on a 3.37-year cycle in catch. The possibilities of time-series relationships, however, were not fully appreciated until Ricker (1954) studied the periodicity of the reproduction function.

The nature of the time-series dynamics may be appreciated by following the Beverton-and-Holt discussion (see also Ricker, 1954). The dynamics of the recruitment-stock relation is linked to the interaction of the α-function and the γ-function. The α-function represents the recruitment-stock relation or the juvenile-mortality transition usually presented in the literature; the γ-function represents the egg transition. (Both functions were mentioned in Chapter 4.) The γ-function, a special case of which Ricker (1954) called the replacement line, expresses the relation of the number of eggs to the number of recruits. In the context of recruitment-stock theory, the literature has had little to say about replacement (see, however, Garrod, 1973, and Garrod and Knights, 1979) or about the form of the replacement line, which is almost universally portrayed as a straight line (see, however, Jones, 1973). Typical dynamic behaviors of an α-function and a γ-function are shown in Figure 5.8.

Nevertheless, it is quite conceivable that the replacement line is not linear but curvilinear. Curvilinearity would result if egg production per unit stock varied as a result of stock density. (Evidence for a curvilinear "replacement line" is discussed in Chapter 6.) As seen from (4.1), the usual depiction of the recruitment-stock relation in terms of *only* the α-function cannot fully depict the dynamics of the stock.

In actuality, the study of different forms of α- and γ-functions permits a broader appreciation of the recruitment-stock dynamic process. To study, for example, the effect of density dependence, Beverton and Holt considered straight-line, or density-independent, α- and γ-functions,

$$R = \alpha S, \qquad S = \gamma R. \tag{5.39}$$

These show that when the stock is in equilibrium (that is, when $R = S$), the α- and γ-functions are coincident (Figure 5.9). But a displacement of either will result in the stock's exploding to infinity or becoming extinct (for example, Figure 5.8 shows an "explosion"). Since neither of these situations is plausible, the relationships (5.39) are jointly untenable. Thus for any stock displaced from equilibrium to return to equilibrium, one of the equations in (5.39) must be curvilinear. The usual assumption is that

Figure 5.8 The dynamics of a self-regenerating model with a density-independent larval mortality rate. In part (*a*) the alpha and gamma functions are equivalent, and the stock is in equilibrium. In part (*b*) the gamma function is greater than the alpha function, and the stock collapses. In part (*c*) the alpha function is greater than the gamma function, and the stock explodes. (From Beverton and Holt, 1957: fig. 6.1.)

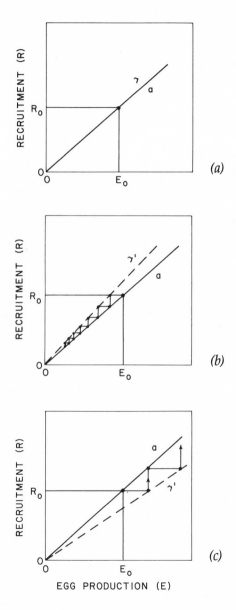

Figure 5.9 The dynamics of a self-regenerating model with the egg-recruit relationship. As may be deduced by the angle of intersection between the alpha and gamma functions, part (*a*) shows that the recruitment-stock dynamics converges on a stable point; part (*b*) shows stable oscillations, and part (*c*) shows unstable oscillatory behavior. (From Beverton and Holt, 1957: fig. 6.6.)

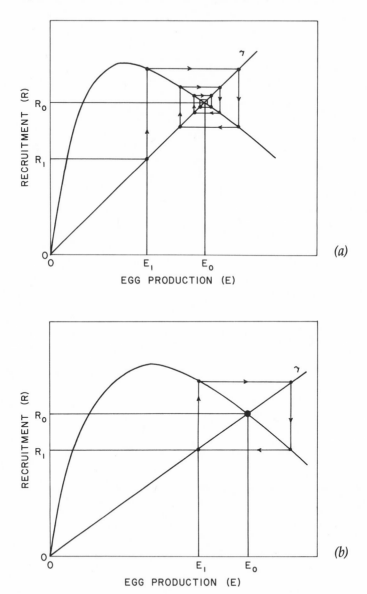

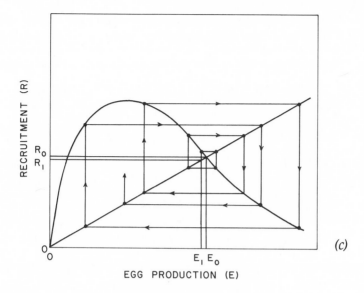

(c)

the first equation, which expresses recruitment as a function of stock size, is curvilinear.

There are three possible dynamic behaviors for a curvilinear α-function and a straight-line γ-function. These behaviors are defined in terms of the angle of intersection between the α-function and the γ-function (Figure 5.9). In the first case, where the angle is less than 90°, the stock proceeds from its displaced value toward the equilibrium value, with a continually dampened periodicity. In the second case, where the angle is 90°, the stock does not proceed toward the equilibrium value; it enters a stable-cycle mode. In the third case, where the angle is greater than 90°, the stock diverges from the equilibrium value.

Note that any curvilinearity in the γ-function can contribute to the stability of the stock in the sense that curvilinearity affects the angle of intersection between the α- and γ-functions. There are two important points. First, under some conditions a curvilinear γ-function can hasten the return to equilibrium. Second, a stable-population behavior *can* be obtained with a linear α-function, provided that the γ-function is curvilinear (this observation can explain the paradox of an empirical linear α-function; cf. Garcia, 1983).

The correlations that may exist among years would naturally be analyzed by using time-series analysis. If the time series can be reasonably stationary or transformed to be stationary, then two time-series models might be considered. The first is the autoregressive model, for example,

$$(x_t - \mu) = \alpha_1(x_{t-1} - \mu) + \alpha_2(x_{t-2} - \mu). \tag{5.40}$$

In this case x_t could be thought of as a deviation from average recruitment in year t, and x_{t-1} and x_{t-2} the deviations in the two previous years. If $\alpha_i < 0$, then the system would tend to be self-correcting.

The other possibility is the moving-average model,

$$(x_t - \mu) = \beta_0 z_t + \beta_1 z_{t-1} + \beta z_{t-2}. \tag{5.41}$$

Here $x_t - \mu$ would be a function of the random variables z_i, where each z_i could be a function of the number of individuals contributing to the biomass of the t, $t - 1$, . . . age groups of the spawning populations, and β_i would reflect the magnitude of transfer of recruits for each z_i.

The time-series models are flexible because they can be written in a variety of ways. For example, (5.40) and (5.41) can be combined by using different terms to obtain the mixed autoregressive moving-average model. Yet at the same time, the models seem inflexible because in the most simple and practicable form they are difficult to reconcile with what seem to be underlying recruitment-related phenomena. "Good fits" might be obtained from more complex forms of the models, but there are numerous questions regarding their interpretation, even though the models suggest interesting phenomena, particularly when transformed to Fourier or frequency space (see, for example, Steele and Henderson, 1984).

The analysis in frequency space might suggest cyclic behavior reminiscent of the periodicities or apparent periodicities that can be generated from Leslie matrices. The difficulty, however, is to accept that, given a ten-year cycle, the complex system actually behaves in such a way as to produce such a cycle. How does the complex system manage to maintain long-term periodicities? After all, events in Fourier space can be transformed to time space; in time space the autocorrelations that generate the frequency responses should be particularly evident and explicable.

A simpler and more illustrative approach might involve a nonparametric representation of the time series. There are basically two approaches: one considers the time series of recruitment and the other considers the time series of recruitment and stock. With respect to the analysis of the recruitment time series, consider

$$x_1, x_2, \ldots, x_n, \tag{5.42}$$

where x_i is the estimate of recruitment in the ith year, and n is the number of years for which data are available. There are $n - 1$ intervals between adjacent recruitment values. If $x_{i+1} > x_i$, the interval is assigned a positive value (+); if $x_{i+1} \leq x_i$, the interval is assigned a negative value (−). Thus each recruitment time series can be represented by a sequence of plus and minus signs. The question arises, then, as to the likelihood that a particular sequence of plus and minus signs reflects a recruitment pattern in which large year classes and small year classes occur independently, or in which large year classes are likely to be followed by large year classes and small year classes followed by small year classes (the discussion of this material follows an analysis by Rothschild and Brunenmeister, 1984).

For example, consider a population in which recruitment has been estimated for seven successive years ($n = 7$). We obtain the sequence +, −, −, +, −, +, which we decompose into "runs"—that is, subsequences that contain elements of the same sign. The sequence +, −, −, +, −, + thus consists of five runs: four runs of length 1 (+, +, −, +) and one run of length 2 (−, −). Table 5.1 shows the possible arrangements of three plus and three minus signs classified by the number of runs, and also the probability distribution of the number of runs. Note that if the plus signs were randomly distributed among the minus signs, the modal value of the number of runs would equal 4. Also, if the distribution of plus and minus signs was random, the sequence +, +, +, −, −, − would be just as unusual as the sequence +, −, +, −, +, −.

Although probabilities can be determined simply by tabulating sequences of plus and minus signs as in Table 5.1, it is more convenient to use the formula developed by Feller (1957) to obtain the probability of the number of runs (k), given the number of plus signs (r_1) and minus signs (r_2), when $k = 2v$ is even:

$$P_{2v} = 2 \binom{r_1 - 1}{v - 1}\binom{r_2 - 1}{v - 1} \bigg/ \binom{r_1 + r_2}{r_1},$$ (5.43)

or when $k = 2v + 1$ is odd:

$$P_{2v+1} = \binom{r_1 - 1}{v}\binom{r_2 - 1}{v - 1} + \binom{r_1 - 1}{v - 1}\binom{r_2 - 1}{v} \bigg/ \binom{r_1 + r_2}{r_1}.$$ (5.44)

Feller (1957) also gives a formula for calculating the most probable number of runs (k), given the number of plus (r_1) and minus (r_2) signs:

$$\frac{2r_1 r_2}{r_1 + r_2} < k < \frac{2r_1 r_2}{r_1 + r_2} + 3.$$ (5.45)

Table 5.1 Possible time series of recruit variations from a hypothetical population for which
recruitment was measured for seven years
and for which the numbers of increases and
decreases in recruitment strength were equal.

Total number of runs	Possible time series	Probability of total number of runs
2	+++−−−	0.10
	−−−+++	
3	+−−−++	0.20
	++−−−+	
	−+++−−	
	−−+++−	
4	+−++−−	0.40
	++−+−−	
	−−++−+	
	+−−++−	
	−+−−++	
	++−−+−	
	−++−−+	
	−−+−++	
5	+−−+−+	0.20
	−++−+−	
	+−+−−+	
	−+−++−	
6	+−+−+−	0.10
	−+−+−+	

To apply this methodology to actual data, we utilize recruitment time
series conveniently tabulated for twelve stocks of fish by Garrod (1982).
For each stock, Table 5.2 lists the length of the recruitment time series, the
observed number of runs, the most probable number of runs, the probability of the observed number of runs, and the probability of a greater
number of runs than that observed.

We can see from Table 5.2 that the number of runs in the recruitment
time series of the Norwegian spring-spawning herring, the North Sea
herring, the St. Lawrence mackerel, the North Sea haddock (for the years
1920–1936), and the Georges Bank haddock is not consonant with the
hypothesis that the plus signs are randomly distributed among the minus
signs, in that for each stock the probability of obtaining more runs than

Table 5.2 Runs analysis of recruitment series for various fish stocks.

Stock	Range containing most probable number of runs	Length of series (years)	Observed number of runs	Probability of observed number of runs	Cumulative probability of greater number of runs than observed
1. California sardine	15–18	31	14	0.116	0.825
2. Norwegian spring-spawning herring[a]	8–11	20	12	0.058	0.092
3. North Sea herring[a]	11–14	23	15	0.059	0.097
4. St. Lawrence mackerel[a]	4–7	10	8	0.063	0.071
5. Arcto-Norwegian cod	12–15	26	12	0.146	0.777
6. Northeast arctic haddock	12–15	26	14	0.164	0.498
7. Greenland cod[a]	9–12	20	13	0.089	0.166
8. St. Lawrence cod[a]	12–15	25	14	0.155	0.407
9. North Sea haddock					
a. 1920–1936	8–11	17	13	0.010	0.010
b. 1946–1976	15–18	31	18	0.114	0.282
10. Georges Bank haddock	13–16	27	18	0.044	0.078
11. North Sea plaice	14–17	30	14	0.133	0.775
12. North Sea sole	9–12	22	13	0.089	0.166

Source: Recruitment data from Garrod, 1982.
[a] Stocks evidencing overall downward trends in recruitment.

observed was less than 0.10. Figure 5.10 shows that recruitment sequences for most species have more runs than would be expected if positive recruitments were randomly distributed among negative recruitments.

Hence there is weak statistical evidence (results are "significant" at the 10 percent level as opposed to the 5 or 1 percent level) that sequences of fluctuations are not random in the recruitment of several fish populations for which Garrod (1982) presents recruitment statistics. The nature of these distributions is unusual for a random distribution because the increases and decreases in recruitment are more evenly distributed than would be expected. These distributions may reflect compensation, as an increase in recruitment tends to be soon followed by a decrease in recruitment, and vice versa. In fact, the magnitude of the cumulative prob-

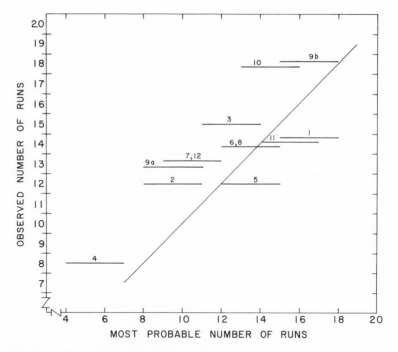

Figure 5.10   The actual number of runs plotted against the most probable number of runs as computed by equation (5.45) for a variety of stocks. The numbers within the figure identify the stocks as listed in Table 5.2. Note that the observed numbers of runs for stocks 10, 3, 7, 12, 2, and 4 are greater than expected. Those for stocks 9b, 6, 8, 11, and 5 are within the range of expectation. Only stock 1 shows an aggregation of plus and minus signs that is greater than random.

ability that a greater number of runs will occur than that observed is an index of this compensatory effect. This index enables us to rank each stock according to degree of compensatory behavior, from greatest to least:

1.	North Sea haddock, 1920–1936	0.010
2.	St. Lawrence mackerel	0.071
3.	Georges Bank haddock	0.078
4.	Norwegian spring-spawning herring	0.092
5.	North Sea herring	0.097
6.	Greenland cod	0.166
7.	North Sea sole	0.166
8.	North Sea haddock, 1946–1976	0.282
9.	St. Lawrence cod	0.407
10.	Northeast arctic haddock	0.498
11.	North Sea plaice	0.775
12.	Arcto-Norwegian cod	0.777
13.	California sardine	0.825

A better understanding of the dynamics of these stocks would require investigation into the mechanisms that produce the interannual linkages in the North Sea haddock, the St. Lawrence mackerel, the Georges Bank haddock, the Norwegian spring-spawning herring, and the North Sea herring, as well as investigation into factors that might cause these linkages to deteriorate.

Considering *both* the recruitment time series and the stock time series, Rothschild and Mullen (1985) classified each year as S_1, low stock/low recruitment; S_2, low stock/high recruitment; S_3, high stock/high recruitment; or S_4, high stock/low recruitment. The time sequence could then be represented by element S_{ij} where $i = 1, \ldots, 4$ classification states and $j = 1, \ldots, n$ years. They found random patterns in some stocks and nonrandom patterns in others.

Multiple Life-History Stages

As suggested in Chapter 4, low-fecundity population-dynamics theory is only marginally adequate for addressing problems associated with the dynamics of high-fecundity populations. Responding to the inadequacy of low-fecundity population-dynamics theory, fishery biologists have developed the recruitment-stock theory, which focuses on the life-history stages neglected by low-fecundity population dynamics.

The theory, despite its importance, presents something of a paradox. On the one hand, there must be a relation between recruitment and parent stock; otherwise there would be no population-dynamics process, and without this process, the abundance of populations would not ex-hibit long-term stability. On the other hand, empirical observations of stock and recruitment do not seem to correspond to the theory.

There is, of course, the possibility that the density of spawners has little relationship to reproductive success and that other sorts of regulatory mechanisms operate. Yet it is possible that alternative configurations of the recruitment-stock problem might reflect a clearer understanding of the interaction of recruitment and stock and its relation to the environ-ment.

An exploration of the problem strongly indicates that researchers have not been efficient in identifying the sources of the large variation often associated with empirical recruitment-stock data (see Figure 5.2). In ad-dition, the traditional recruitment-stock problem structure may very well mask important elements of the population-dynamics process.

Remedying these problems would involve examining the components of variation in recruitment-stock data. There would be a tendency to initiate such a search by considering the effect of the "environment" on stock and recruitment, as it is often stated that deviations from the re-cruitment-stock curve are caused by environmental variation. But the importance of environmentally induced variation cannot be assessed without further analysis of the density-dependent effect of stock and recruitment.

Insights into recruitment-stock variation may be obtained by consider-ing the contribution of life-history stages to this variation. Paulik (1973) suggested several graphic methods to facilitate the analysis of multiple life-history stages. He based his thinking on Johnson's (1965) division of the life history of the sockeye salmon into the following stages: (a) spawning → fry, (b) lake residence, (c) migration to sea, and (d) return to spawn. Paulik also cited the work of Larkin, Raleigh, and Wilimovsky (1964) and Junge (1966) on two-stage life histories that involved com-pensatory and depensatory interactions, as well as his own work (Paulik and Greenough, 1966) on salmon management.

Paulik pointed out that Equation (5.3) could be written in the form

$$R = f_3\{f_2[f_1(S)]\}. \tag{5.46}$$

Hence f_1 might be the relation of the biomass of spawning fish to egg production; f_2 might be the relation of egg production to the number of larvae that hatch from the eggs; and f_3 might be the relation of the number of larvae to the number of recruits.

The dynamic behavior of a multiple-life-history-stage model can be assessed by examining the coordinate system of a Paulik diagram (Figure 5.11). The system has four quadrants: quadrant 1 contains the recruitment-stock relationship, or the relation of the number of recruits to the

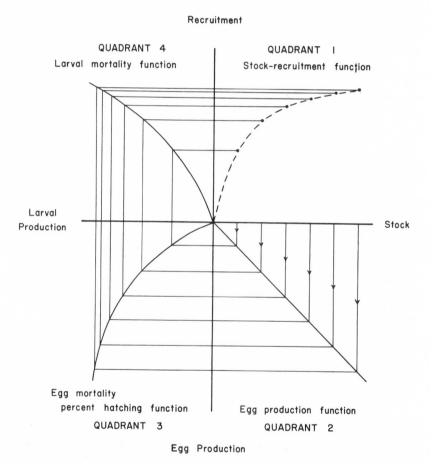

Figure 5.11 A Paulik diagram showing the construction of a recruitment-stock curve. Each quadrant contains a function representing a particular transformation in the recruitment-stock process. This example shows a linear density-independent stage in quadrant 2 and density-dependent stages in quadrants 3 and 4. Note that the diagram is not drawn to scale.

biomass of spawning fish; quadrant 2, the relation of the biomass of spawning fish to egg production; quadrant 3, the relation of the number of eggs to the number of larvae; and quadrant 4, the relation of the number of larvae to the number of recruits.

The Paulik diagram shows how transitions between life-history stages affect the recruitment-stock relation. The diagram in fact points out the importance of the among-stage transitions. These transitions may be categorized as density independent, density dependent, or strongly density dependent. A density-independent transition may be represented as a directly proportional transfer from one stage to the next. A density-de-

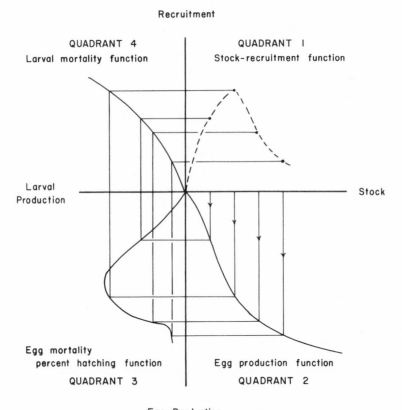

Figure 5.12 A Paulik diagram showing the generation of a dome-shaped recruitment-stock function with a dome-shaped function in quadrant 3. Note that in contrast to Figure 5.11 the function in quadrant 2 is curvilinear. The diagram is not drawn to scale.

pendent transition is represented by a curvilinear relation from one stage to the next. Transitions that are simply density dependent are represented by an asymptotic function, whereas transitions that are strongly density dependent are represented by a dome-shaped function.

For example, Figure 5.11 shows the construction of a recruitment-stock curve where there are two density-independent stages and one density-dependent stage. This construction clearly generates an asymptotic recruitment-stock curve in the first quadrant. As another example, Figure 5.12 shows the construction of a recruitment-stock curve with a strong density-dependent function; it can be seen that this construction generates a dome-shaped curve.

The multiple-life-history-stage presentation then reflects the possible density-dependent operation of several quasi-independent life-history stages, which, when taken collectively, constitute a "fail-safe" series of regulatory mechanisms. If one fails, then the other may generate a more intense effect. This model suggests that explosions and collapses occur in the relatively unlikely event that the regulatory mechanisms in the system break down, as reflected by a breakdown at each life-history stage. Clearly, the probability of dysfunction of regulatory mechanisms at all stages is a function of the number of stages in any particular life history (stage is not necessarily morphological — rather, it is defined somewhat abstractly as a quasi-independent unit having its own density-dependent characteristics). Hence those organisms with a greater number of stages belong, in principle, to more stable populations than do organisms with fewer stages, unless a greater number of stages increases the probability that any one stage would be more susceptible to the challenge of environmental aberrancies — a question of particular interest to those concerned with anthropogenic effects, whether resulting from habitat modification or from fishing.

Conclusion

This chapter has explored the seeming paradox of recruitment-stock theory. On the one hand, long-term population stability suggests that we should expect a strong relation of stock to recruitment; on the other hand, empirical evidence suggests that the relation is vague in that variances about the theoretical relationship seem fairly large.

Accordingly, we are given to conclude that (a) the relationship is indeed not very precise, but it works sufficiently well to regulate the population; (b) the relationship is relatively precise, but its precision is masked by varying conditions or environmental domains; or (c) any relationship that might exist is occluded by measurement error.

In pursuit of the appropriate conclusion we have examined some nuances of the recruitment-stock relationship. Our examination reflects that the apparent recruitment-stock paradox revolves about two issues. The first is the variability in the relationship, and the second is its curvilinearity. The applicability of the relationship has been challenged because of the large variability in empirical recruitment-stock data. We can now see that low variability is not a particular requirement of the recruitment-stock relationship, while curvilinearity is. In fact, the high variability may reflect the capability of the reproductive process to accommodate widely varying conditions. Recruitment-stock functions such as (5.4) and (5.5) mask the complexity of the process and hence make it difficult to weigh the relative importance of the factors intermediate between parent stock and subsequent recruitment in contributing to the variability or the curvilinearity in either the γ-function or the α-function.

In order to examine the nature of the form of the relationship and its variability, I have decomposed the recruitment-stock relation into component parts. Remarkably one part, egg production, is largely a property of the fish, whereas another part, egg-and-larval mortality, is largely a property of the oceanic environment. Put another way, the events associated with quadrant 2 are at least proximally related to egg production, while events associated with quadrants 3 and 4 are at least proximally related to properties of the oceanic environment. Functions in each quadrant seem to be quite independent of one another in that the "environment" of the relatively long-lived, egg-producing, nektonic adult is quite different from the "environment" to which the planktonic eggs and larvae are exposed during their brief existence.

The next two chapters seek to find a basis for variation in the recruitment-stock relation by considering quadrant-2 problems, the production of eggs, and quadrant-3 and -4 problems, the fate of eggs and larvae in the sea.

6

Egg Production

Variability in egg production appears to be an important component of the population-dynamics process. The variability seems to be related to female nutritional history and thus to population abundance. At depressed levels of population abundance, each female has more food, improved nutrition, and increased growth. In addition, egg production per spawning biomass increases, as well as egg survival rate, owing to improved "egg quality." As Nikolskii (1962: 269) said, "Many examples are known when a change in food supply is associated not only with a change in growth and the time of maturity but also with an alteration of the fecundity in fish of the same size." He also observed that "improvement in food supply leads to an increase in fecundity in individuals of the same size. This fecundity increase can be reached either by an increase in the volume of the gonads or by a decrease in egg size," referring specifically to studies on Sakhalin herring *(Clupea harengus pallasi)*, North Sea herring, north Caspian roach, and north Caspian common carp.

Parts of this scenario appear to be well supported. The most verifiable inferences suggest that individuals in numerically depressed populations grow faster and mature earlier than individuals in more abundant populations. For example, Jakobsson and Halldorsson (1984) provide evidence of increased growth and reduced age at maturity in a numerically depressed stock. In the early 1960s the Icelandic spring-spawning herring population was at a peak of abundance, about 2500×10^6 fish; in the mid-1970s the population declined to a low of about 100×10^6 fish; but by the late 1970s its abundance increased to nearly the former peak level. When the population was at its lowest level of abundance, the mean length and weight of the fish were greatest, and the percentage of mature three-ring and particularly two-ring fish increased substantially (Figures 6.1 and 6.2). When the population was at a relatively high level of abundance, condition factors for both young and old fish declined. The increase in growth and decrease in age at maturity caused a substantial increase in spawning stock per fish.

Whereas the dynamic behavior of the Icelandic summer-spawning herring follows the textbook pattern, that of other populations appar-

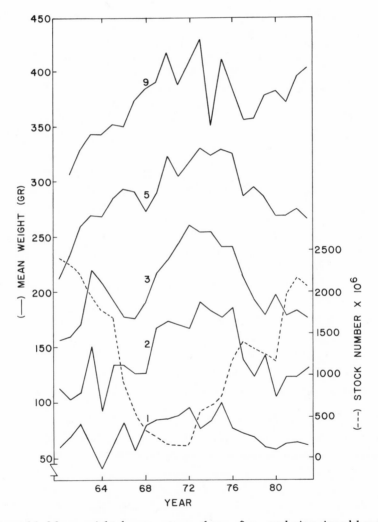

Figure 6.1 Mean weight for one-, two-, three-, five-, and nine-ringed herring from 1960 to 1983. Total stock size in number is shown as a dashed line. (Jakobsson and Halldorsson, 1984: fig. 1, reprinted with permission of the authors.)

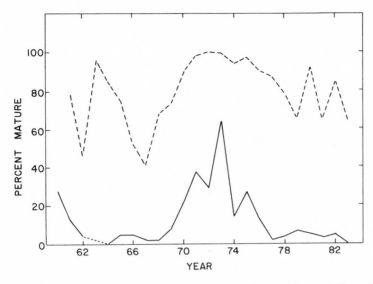

Figure 6.2 Maturity rate for two- and three-ringed herring from 1960 to 1983. The dashed line refers to three-ringers, and the solid line to two-ringers. (Jakobsson and Halldorsson, 1984: fig. 4, reprinted with permission of the authors.)

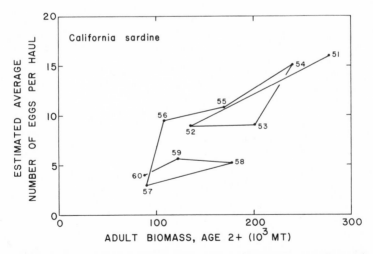

Figure 6.3 Relation between estimated average number of sardine eggs per haul and adult biomass in the Pacific sardine from 1951 to 1960. (Data based on Zwiefel, 1973, and MacCall, 1979.)

ently does not. It is interesting to compare egg production relative to its biomass in the Pacific sardine, for example. MacCall (1979) revised Murphy's (1966) estimates of California sardine biomass. These estimates can be compared with Zweifel's (1973) estimates of the average number of sardine eggs per haul (Figure 6.3). Figure 6.3 suggests the existence of two distinct egg-production/biomass regimes, one from 1951 to 1956 and the other from 1957 to 1960. Could the reduction of roughly 50 percent of egg production per unit of spawning biomass be a function of reduced stock abundance (which is contrary to expectations), or is it a function of environmental change?

These examples set the stage for the study of egg production. On the one hand, fish seem to respond to population reduction by increasing their egg production. On the other hand, sometimes the expected density-dependent response does not operate. This chapter considers two parts of the egg-production problem: the elaboration of biomass and the nature of egg superabundance.

Elaboration of Biomass

Cohort biomass is a function of the initial number of individuals in the cohort, their growth, and their mortality. The mature-cohort biomass consists of somatic and reproductive components. These components are not totally independent; egg production is limited by the female's somatic bulk, and portions of somatic biomass are mobilized to contribute to reproductive biomass. The parallel elaboration of somatic and reproductive biomass is considered in terms of both growth and mortality.

Growth

Fish growth is generally estimated from measurements of length taken at different ages. The resulting growth-in-length curve is not always satisfactory, as interest is often greater in growth-in-weight curves. To obtain the latter, it is common practice to transform lengths-at-age to weights-at-age, assuming that weight is a cubic function of length, which is probably true only mathematically or under particular "average circumstances." Growth in length and its cubic approximation to weight is usually satisfactory for stock-assessment analyses but not for detecting seasonal increments and decrements in somatic and reproductive biomass. To detect these changes, actual weight measurements are needed.

Various complications are attendant to the study of growth in weight, even if actual weight measurements are available. In the study of the growth of a particular organ system, periodicities in organ-system weight can occur at frequencies greater than periodicities in the total weight of the organism. Hjort (1914) assembled some of the first observations on growth periodicity in fat bodies or "ister" of herring, sprat, and pilchard, as well as in cod liver. He noted that ister was not present in young fish or in mature fish and that it increased in the fall and disappeared in the winter: "The supply of fat increases during the summer and is consumed during the winter, while water is excreted in the summer and assimilated in the winter. During the winter, part of the dry matter in the system is consumed, and replaced by water, so that no great loss in weight is apparent. The quality of the fish, however, is considerably affected" (1914: 171). Thus, owing to possible changes in water content of the fish, the weight metric may be a somewhat confounded index of somatic and reproductive growth. The problem of water content was discussed later by Iles and Wood (1965) and Iles (1965, 1984). They showed that in North Sea herring, water content could vary between 60 and 75 percent, depending on the season, so any use of the weight metric for growth studies would need to take into account consistent seasonal periodicities in water balance.

Periodicities in organ-system growth are well documented for some systems, but poorly documented for others (Iles, 1984, also discusses seasonal periodicities in the percentage of biomass of reproductive and somatic tissue, as well as the protein-lipid ratio of the latter). Periodicities in reproductive-system growth, for example, are fairly well known: the growth cycle is quiescent in immature fish and is rapid, usually annual, in mature fish. The growth of the hepatic system in fish appears to be quite variable. Hjort commented on the weight of liver per unit weight of mature cod or skrei. Skrei livers could vary between 10 g and 1 to 2 kg and could contain about 20 to 60 percent oil. Interannual fluctuation in the quantity of liver per unit weight of skrei and its close correlate, the percentage of fat in the livers, seems considerable. Hepatic growth is not nearly as well studied as reproductive growth, but it evidently plays an important role in the storage of energy, which is mobilized into reproductive tissue, as implied by de Vlaming (1971) and Tyler and Dunn (1976) (see Figure 6.4).

Growth curves (for example, the von Bertalanffy and Pütter curves, Equations [1.7] and [1.8]) that do not fully take into account reproductive growth, either because of their structure or because their parameters are estimated so as to diminish the importance of reproductive growth, give a misleading impression of individual or population elaboration of bio-

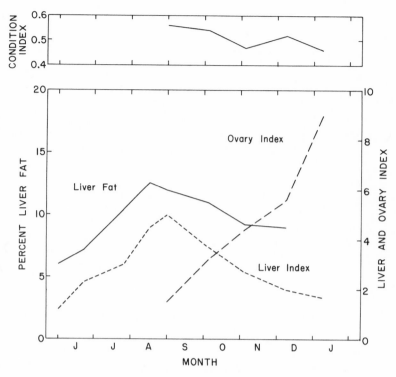

Figure 6.4 Monthly trends in mean values of condition index, ovary index, liver index, and percentage of fat in liver for winter flounder *(Pseudopleuronectes americanus)* in Passamaquoddy Bay. (From Tyler and Dunn, 1976: fig. 12.)

mass, particularly in terms of energetic considerations. As traditionally used, growth curves suggest that growth is essentially a phenomenon that occurs mostly in juvenile or immature fish, and that adult fish grow relatively little because proportional weight increments in juvenile fish generally exceed those of adult fish. (In fact, many of the detailed studies of growth are based only on the growth of juvenile fish; see, for example, Brett, 1979.) This conclusion is, however, based on an artifice, because there is considerable adult growth in reproductive tissue.

Note that most adult growth is not in somatic tissue but in reproductive tissue, which tends to be "averaged out" from weight-at-age calculations because of its seasonal periodicity. Using de Veen's (1976) data on age-specific somatic and reproductive biomass for the North Sea sole, Figure 6.5 shows growth curves based on (a) minimum somatic weight at age and (b) minimum somatic weight at age plus maximum annual accumulated ovary weight. The cumulative growth curve seems to represent elaboration of biomass in a female fish more accurately than the average growth curve, yet in virtually all analyses the average growth curve is purported to represent growth. As empirical data tend to be based more on the average-growth-curve model than on the cumulative-growth-curve model (see Ricker, 1979), it is little wonder that attempts to establish a theoretical basis for mature-fish growth have not been satisfactory.

The problem may be better analyzed by considering separately immature- and mature-fish growth. The study of immature growth involves somatic growth and the factors that cause the onset of maturity. The study of mature growth involves understanding the interactions between reproductive and somatic growth as parallel processes.

Growth of immature fish. The study of somatic growth in immature fish is relatively straightforward. Food is ingested and allocated to growth, activity, maintenance, and heat (see, for example, Brett and Groves, 1979). The allocation of energy among these functions operates under a complex control system affecting the relationship between percentage of body weight, daily growth, and percentage of body weight, daily ration. Brett (1979) uses a growth-ration curve (G-R) to facilitate the study of the

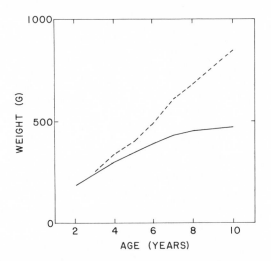

Figure 6.5 Growth curves of the sole based on minimum somatic weight at age *(solid line)* and minimum somatic weight plus maximum annual accumulated ovary weight *(dashed line).* Not taking account of reproductive growth would generate a growth curve close to the solid line; taking account of reproductive growth would generate a growth curve close to the dashed line. (Based on de Veen, 1976.)

interrelations. He identifies three points on the G-R curve: the zero-growth, maintenance-ration point; the optimal-growth, optimal-ration point; and the maximum-growth, maximum-ration point. He then reviews what is known about the interrelationship of these points relative to both abiotic and biotic factors. If we apply inferences from Brett's work to interannual differences in oceanic growth of juvenile fish, we find that abiotic factors would have relatively minimal effect and ration would have a strong effect. In other words, the average effects of abiotic factors over the range of values as they might exist in the sea seem to be small relative to the average effects of change in food supply.

The inference that food supply affects the growth of juvenile fish in the sea is based on observations such as Brett's and on the fact that juvenile fish belonging to large year classes seem to grow more slowly than those belonging to small year classes.

The literature reflects various interpretations of density-dependent growth in immature fish. For example, according to Ware (1980: 1013–14), "during the immature phase of the life history, the average growth rate of many fishes is density dependent. This phenomenon has been widely observed in Atlantic cod *(Gadus morhua)* (Crabtree and Ware, 1975; Cushing and Horwood, 1977); haddock *(Melanogrammus aeglefinus)* (Raitt, 1939; Templeman, Hodder, and Wells, 1978); Atlantic herring *(Clupea harengus harengus)* (Iles, 1968; Anthony, 1971; Hubold, 1978); Pacific sardine *(Sardinops sagax)* (Iles, 1973); flatfish (*Pleuronectes platessa* and *Solea solea*) (Rauck and Zijlstra, 1978); and American shad *(Alosa sapidissima)* (Leggett 1977)." Iles (1968) cites both examples where density-dependent immature growth has been claimed to operate and examples where it is believed not to operate (Hjort, 1932; Bowman, 1932; Walford and Mosher, 1943; Marr, 1960, 1963).

The nature of the density-dependence argument will be evident from a few examples. Raitt (1939) was one of the first to report density-dependent growth, and his work is frequently cited in support of the density-dependent growth phenomenon. He examined North Sea haddock growth from 1928 to 1936, providing data on the relation between average length of fish at various ages and population density. Figure 6.6 shows some of Raitt's data selected to show the relation of length of two-year-old haddock to an index of recruitment. The figure can be interpreted in various ways. One is that there is a density-dependent relationship that can be fit through the points (that is, growth on the average is reduced at higher stock sizes); another is that there is simply considerable variability in mean size when the stock is at a low level of abundance; and a third is that two sets of conditions prevailed when the

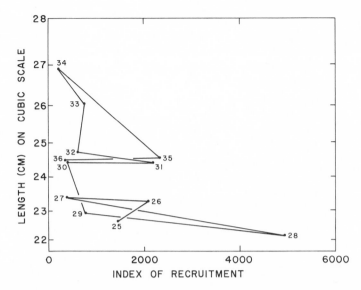

Figure 6.6 Average length attained by various year classes of North Sea (Area C) haddock by the end of their second year of life as a function of an index of recruitment. Lengths are plotted on a cubic scale to emphasize change in weight. Note that the impression of density-dependent growth is dominated by the growth in a single year class, 1928. (Data from Raitt, 1939.)

stock was at a relatively low level of abundance: from 1925 to 1929 conditions for growth were poor, and from 1930 to 1936 conditions for growth were good.

Iles (1967, 1968, 1970) studied the growth of East Anglian herring. The young herring spend their first year of life in inshore waters and their second in an offshore nursery ground. In the third year of life individuals larger than 19.5 cm in length become mature and recruit to the fishery. If, however, an individual does not reach 19.5 cm by its third year, it does not become mature and does not recruit to the fishery until its fourth year. Figure 6.7 shows mean lengths of herring derived by Iles. Rates of growth were low during the World War II years, when the population was at a high level of abundance. When fishing was resumed after the war, the stocks were reduced in abundance and the growth rate increased. As pointed out by Iles and discussed by Burd and Cushing (1962), however, 1950–1955 were years of relatively high *Calanus* abundance, an important food of the herring. Although reduced stock levels could stimulate increased growth, so could an increase in *Calanus*, an apparently *density-independent* event. Figure 6.7 also shows the per-

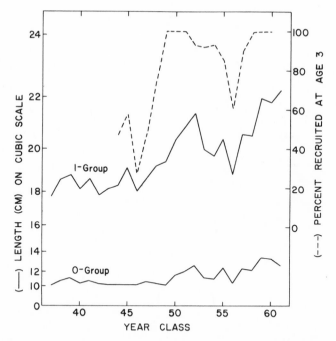

Figure 6.7 Lengths of 0-group and I-group East Anglian herring and percentage of 3-year-old recruits. Lengths are plotted on a cubic scale to emphasize changes in weight. Note increasing growth of I-group fish and maturing of younger fish at the resumption of fishing immediately after World War II. (Based on Iles, 1967, 1968.)

centage of three-year-old fish that mature in each year class, reflecting the fact that percentage of maturity is correlated with growth rate. The implication is that size at maturity is constant, and hence faster-growing fish reach the size of maturity before slower-growing fish, which is distinct from the possibility that size at maturity is reduced in faster-growing fish.

Thus, notwithstanding some examples to the contrary, as a general rule density-dependent growth appears to occur in immature fish. It also seems that the biomass of forage appropriate for immature fish sets in motion the density-dependent processes, because when there are more immature fish, there is less food per fish and so growth is reduced. This inference suggests that there is generally a limited pool of forage for immature fish — which is surprising, since it might be thought that immature fish would not exert an intense mortality on their prey, particularly when the prey is likely to consist of numerous species. But the situation is not so surprising if the density-dependent effects on forage operate only at *extremely* high levels of predator abundance, as seems to be the case in the Raitt example. It appears from the many observations relating growth to year-class size that food is the major controlling element, and environmental factors have relatively little effect on immature growth.

From the prevalence of density-dependent growth in juvenile fish, we might think that the density of juvenile fish is the primary control in the abundance of their forage. But it is obvious that, at least in principle, variations in abundance of forage also affect juvenile-fish growth. There are only a few examples, such as the herring-*Calanus* one, to support this statement — perhaps because forage usually consists of many species, and changes in the abundance of one species would tend not to have a large effect on the total forage. Environmental changes, however, could affect the total forage mix or the consumption of all prey species in concert (for example, an increase in turbidity might reduce predation efficiency).

Effects of immature growth on reproductive potential. Having considered the question of immature or juvenile growth in its own right, we can now consider the relation of magnitude of juvenile growth to egg production. A first thought suggests that growth occurring before maturity could not affect the production of eggs. But this does not seem to be the case. Indeed, the magnitude of immature growth appears to influence an individual female's reproductive potential. Immature fish that are faster growing seem to mature at an earlier age or smaller size than immature fish that are slower growing. In other words, immature females in populations that are growing faster, hence populations that are relatively depressed numerically, mature at an earlier age or smaller size than more slowly growing immature females (thus food resources of juvenile fish are implicated in the age attained at maturity, and, as suggested previously, these food resources are most likely to be affected by the density of the juvenile fish).

Several studies demonstrate the effect of increased juvenile growth on the onset of early maturity. The Jakobsson and Halldorsson and the Iles studies have already been mentioned. Shelton and Armstrong (1983) have shown that mean length at maturity of the South African pilchard (*Sardinops ocellata*) declined 3 cm from a period when the stock was relatively abundant to a period when it was relatively depressed. By contrast, in the South African anchovy (*Engraulis capensis*), which is found in roughly the same area as the pilchard, an increase in population abundance caused a 1-cm increase in mean length at maturity (Figure

6.8). Daan (1978a) reported on the mean length at maturity of North Sea cod: after World War II they matured at a smaller length than before the war, although the linkages between length at maturity and population size are not certain.

These examples show that immature growth rate can have a profound effect on the egg production of mature females. It can affect not only age at maturity be also size. If size triggers maturity, then fish that mature at

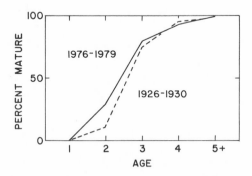

Figure 6.8a Percentage of mature females at age for North Sea haddock. The dashed line is based on Raitt's (1933) data for 1926–1930, and the solid line on data for 1976–1979. (Based on Hislop and Shanks, 1981: table 4. Reprinted with permission of ICES.)

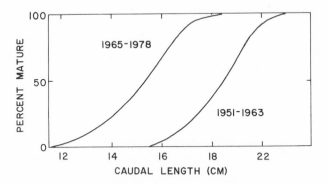

Figure 6.8b Percentage of mature females as a function of length for pilchard sampled from 1951 to 1963 and from 1965 to 1978. During the first period the population was at a relatively low level of abundance; during the second period it was at a relatively high level of abundance. The mean length at 50 percent maturity *decreased* about 3.5 cm. (From Shelton and Armstrong, 1983: fig. 3.)

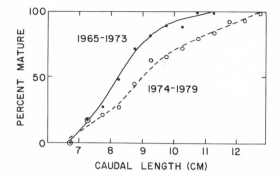

Figure 6.8c Percentage of mature females as a function of length for anchovy sampled from 1965 to 1973 and from 1974 to 1979. During the first period the population was at a relatively low level of abundance; during the second period it was at a relatively high level of abundance. The mean length of 50 percent maturity *increased* about 1 cm. (From Shelton and Armstrong, 1983: fig. 4.)

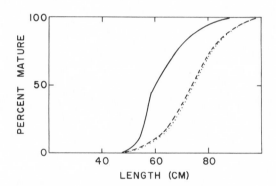

Figure 6.8d Percentage of mature females as a function of length for North Sea cod. The solid curve is based on Holt, 1893, and Graham, 1924; the broken curves are based on Oosthuizen and Daan, 1974. (From Daan, 1978: fig. 51. Reprinted with permission of ICES.)

an early age might not necesssarily be larger than fish that mature at a late age.

Growth of mature fish. At the mature stage the same controlling principles of somatic growth apply as in the immature phase, but they are modified by the fact that anabolism must now be partitioned between reproductive and somatic biomass. Reproductive growth begins with the fundamental germ cells, oocytes. Each oocyte either evolves into an ovulated and spawned egg or is reabsorbed through atretia (see, for example, Vladykov, 1956). The evolution of an oocyte into an egg involves its investment with vitellogen or yolk, a primary source of energy that sustains the egg and the yolk sac larva until it can depend entirely on exogenous sources of food. The process of vitellogenesis is accompanied by deterioration of the germinal vesicle and completion of meiosis, which signals the point at which the egg ovulates and takes on water just prior to spawning.

A setting for the developmental process in herring is provided by Iles (1965, 1984). The elaboration of reproductive biomass is a complex process. Feeding in herring begins at the onset of the production cycle and is curtailed at gonad maturation. Somatic growth begins as soon as feeding begins, but it declines as reproductive growth increases. Somatic fat accrues passively during the feeding period. Fat reserves are then used as an energy source for reproductive growth. Following this model, Figure 6.9 shows that growth in the North Shields herring occurs primarily during May, June, July, August, and September. May, June, and July are

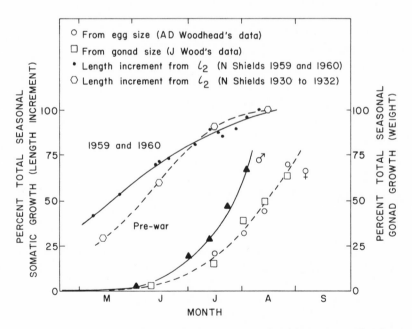

Figure 6.9 Somatic and gonad growth in North Shields herring. The figure shows remarkable periodicity in somatic and reproductive growth. (From Iles, 1965. Reprinted with permission of ICES.)

the months of primary somatic growth, and August and September are the months of primary gonadal growth.

Thus in North Shields herring, feeding conditions in May, June, and July contribute to somatic growth. This growth provides a "physical frame" for reproductive growth, in that larger fish can, among other things, contain more eggs than can smaller fish. It also provides energetic resources, which, in addition to food ingested in August and September, are mobilized into reproductive tissue. Hence as herring feed on organisms that are "close" to primary production, the nature of the productive cycle before spawning can be important in affecting egg production, through both somatic growth and reproductive growth, whereas the nature of oceanic production after spawning affects larval feeding conditions and possibly the extent to which larvae are subject to predation.

In the North Shields example, then, there are three pools of herring food: one that sustains the fish during their somatic growth; one that sustains them during their reproductive growth; one that sustains larval herring after the eggs spawn and hatch. In herring there may be a correlation among the food resources of adults, juveniles, and larvae. Hence, to what extent is population control exerted by a very large biomass of adult herring? In other words, to what extent is the larval-food resource affected by the magnitude of the reproductive-growth food resource, and to what extent is the reproductive-growth food resource affected by predation or the somatic-growth resource? It is clear that a controlling or regulating effect is possible not only when the biomass of herring is large, but also when the biomass of food resource is large.

Production of Eggs

The control of the reproductive process depends on a series of endogenous and exogenous factors. The endogenous factors include a little-understood internal clock, and the exogenous factors evidently involve photoperiod and temperature and the nutritional history of the female (see, for example, Stacey, 1984, and Bye, 1984). These controlling factors appear to be associated with considerable variability in the production and quality of eggs. Egg production—in particular, variability in fish fecundity—was reviewed in detail by Bagenal (1973). He examined interannual differences in fecundity sequences in several species. After taking account of differences in length, he found that the percent difference in the high-fecundity year in each sequence relative to the low-fecundity year was 48 percent for plaice, 40 percent for long rough dab, 28 percent for witch, 25 percent for pike, 56 percent for haddock, 25 percent for Norway pout, and 34 percent for herring—differences that seem quite large.

In a 1966 study Bagenal analyzed the spatial-temporal variability of plaice fecundity throughout its range, using both his own data and those of others (Figure 6.10). Fecundities were compared for plaice whose lengths were adjusted to 37 cm by covariance procedures. Bagenal showed the typical relation between fecundity and length, and exponentially increasing curve exhibiting considerable variability (some of which might be due to the estimation procedure for egg number). He observed both interannual and spatial differences in fecundity. Plaice of the same lengths increased in fecundity by somewhat more than 20,000 eggs, or nearly 15 percent, mostly between 1957 and 1958. With respect to spatial variation, the minimum fecundity, about 100,000 eggs per 37-cm female, was found at the center of plaice abundance, whereas fecundity at the perimeter of the main population had increased to about 160,000 eggs per 37-cm female. This example is particularly interesting because Andrewartha and Birch (1954) supported the idea of *density-independent* population control by pointing out that organisms do *not* exhibit density-dependent responses at the edge of their ranges.

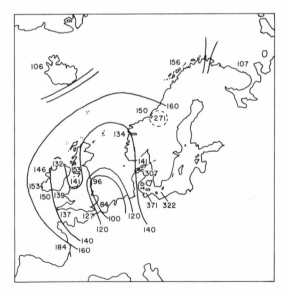

Figure 6.10 Spatial variability in plaice fecundity. Fecundity increases with distance from the center of the range, the southern North Sea. The fecundity data are all adjusted to plaice 37 cm in length. (From Bagenal, 1966: fig. 4.)

Bagenal's observations thus provide examples of temporal-spatial variability of fecundity for fish that spawn only once a year. In species that spawn several times a year, the possibilities for increased sources of variability are considerably enriched. Batch spawners have an opportunity to produce more or fewer eggs by altering not only the number of eggs at each spawning but the number of spawnings as well. Hislop, Robb, and Gauld (1978) studied batch spawning in North Sea haddock *(Melanogrammus aeglefinus)*. They found that on average the fish spawned seventeen batches of eggs over a period of thirty-three days. Working with another species, the whiting *(Merlangius merlangius)*, Hislop (1975) found that these fish spawned batches of eggs for somewhat more than a two-month period. In contrast to the haddock, the number of eggs in each batch declined with time. The mean diameter and dry weight of the eggs decreased during the spawning season, and there were differences in egg diameter and dry weight between the two years in which the experiment was conducted. Hislop also showed that spawning can result in a greater weight loss in fish, particularly in individuals (Hislop, 1975: fig. 4).

A technique for the actual estimation of spawning incidence and batch fecundity was discovered by Hunter and Goldberg (1980). For this technique, based on the "life cycle" of an egg, the critical point is the hydration of the yolk oocytes, signifying the imminence of ovulation. A postovulatory follicle will be a remnant at the site of ovulation of each egg. The morphology of the postovulatory follicle permits one to determine the time since spawning, up to about forty-eight hours. In addition, enumeration of the postovulatory follicles permits one to determine the number of eggs in each spawning batch.

Hunter and Leong (1980) used this technique to study the spawning energetics of the northern anchovy. They found that (a) the average female anchovy weighed 16 g; (b) it spawned twenty times during the year; (c) the ovary wet weight was 0.6 g; (d) the ovary dry weight was 0.17 g; (e) the number of eggs in each spawning batch was 6500; (f) the number of eggs spawned each year was 130,000; and (g) the dry weight of each spawned egg was 0.03 mg.

Suppose the ratio of dry weight to wet weight is represented by the ratio of ovary dry weight to ovary wet weight, which is about 3.4 : 1. The wet weight of eggs at each spawning is 0.03 mg $\times$ 6500 $\times$ 3.5 $\cong$ 0.7 g. Thus the egg mass per spawning is 0.7/(16.4 + 0.7); per year it is roughly 14/17.1, or 7/8 the body weight, which is much higher than the ovary-weight/body-weight ratio in most single-batch spawners.

A final feature of variability in fecundity involves egg size, which can vary among closely related populations. Hempel and Blaxter (1967) (see also Blaxter and Hempel, 1963, and de Ciechomski, 1966) made an extensive study of egg size in herring stocks of northern Europe (Figure 6.11). They found that *within a spawning group* there was a *general* increase in egg weight as a function of maternal size. Eggs of young Norwegian herring were 14 percent lighter in weight than eggs of older fish. Because the quantity of yolk was less in the smaller eggs, the yolk-sac larvae could survive on yolk alone for only thirty-nine days compared to forty-one days in the larger eggs. In addition, the larvae that hatched from the smaller eggs were somewhat smaller than those that hatched from the larger eggs. Hence it seems generally true that large eggs produce larvae that are capable of living longer on their yolk supply; larvae hatching from larger eggs are larger than larvae hatching from smaller eggs; and younger fish in a spawning group often have smaller eggs than do older fish in a spawning group.

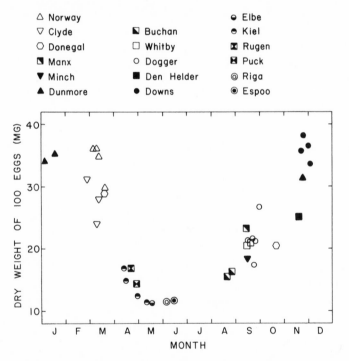

Figure 6.11 Dry weight of herring eggs from various locations in northern Europe showing that large eggs are spawned in the winter months and small eggs are spawned in the summer months. (From Hempel and Blaxter, 1967: fig. 5. Reprinted with permission of ICES.)

Hempel and Blaxter (1967) commented on racial differences in egg size and the temporal-spatial distribution of egg weight. Racial differences can be considerable. For example, the Downs group of herring produces eggs that weigh nearly twice as much as the Buchan group of herring. The question of egg weight in relation to spawning time and place was considered relative to the time of spawning. Hempel and Blaxter (1967: 191 – 192) classify the characteristics of egg production according to the spawning season as follows:

Herring spawning in winter and early spring. Spawning occurs before spring bloom; hence spawning adults utilize body reserves for elaboration of reproductive biomass. Eggs are large and develop at low temperatures, food supplies for young are low; predators on young may be scarce; and larvae hatch with large yolk sacs "on which they grow to a size which enables them to feed on relatively large plankton organisms."

Herring spawning in spring after zooplankton abundance increases. Adult fish have begun to feed before spawning, and larvae hatch under favorable conditions. Eggs are very small. Egg size decreases as the season progresses in the western Baltic either because younger fish are spawning or because of some adaptation to a smaller egg size as the season progresses.

Herring spawning in summer and autumn. Adult fish either interrupt their feeding season in order to spawn or have just finished feeding. Larvae hatch under favorable feeding conditions in July and September. Zooplankton, however, decreases and egg size increases as the season progresses.

So while population fecundity in terms of reproductive biomass seems to be related to nutritional history, the allocation of specific units of egg biomass appears to be related to genetic adaptation (for example, the time, place, and egg size in herring spawning) and environmental control (the number of times batch spawners spawn in a year).

Factors Affecting Fecundity

Bagenal (1973) reviewed possible causes of fecundity changes in fish. He commented on population density, temperature, food supply, and stress, as well as on factors that affect egg size. Again, the nutritional status of the female seems to be an important agent in cohort-specific fecundity.

Experimental studies by Scott (1962), Bagenal (1969), de Vlaming (1971), Wootton (1973), and Hislop, Robb, and Gauld (1978) all implicate the nutritional status of the female in the variability of fecundity. Wootton (1973) studied the effects of food ration on egg production in the three-spined stickleback *(Gasterosteus aculeatus)*. His results were consonant with Scott's (1962) experiments on the rainbow trout *(Salmo gairdneri)*, Bagenal's (1969) on the brown trout *(Salmo trutta)*, and Hester's (1964) on the guppy *(Lebistes reticulatus)*. As Wootton (1973: 95) observed, "the effect of the food supply seems to be primarily on the size at which the female matures. Larger females tend to produce more eggs per spawning, [sic] have more spawnings and to have a shorter inter-spawning interval."

Tyler and Dunn (1976) studied the effects of food ration on the "somatic and organ condition" of winter flounder *(Pseudopleuronectes americanus)*. In a laboratory phase of the study, fish were fed at six different frequencies: 0, 2, 4, 8, 16, and 32 times per thirty-two days for nearly four months. They found that the percentage of fish with yolked eggs, the ovary weight, and the liver weight all increased with increased feeding frequency. They also found that in comparing wild fish with laboratory fish, the percentage of yolkless oocytes declined with the condition factor, again implicating the female's nutritional status as a factor influencing egg production.

In discussing their results, Tyler and Dunn point out that there are two views in the literature relating food supply to fecundity. One view is that low rations reduce fecundity by increasing reabsorption, or oocyte atresia. A second view is that low rations reduce fecundity by reducing the recruitment of new oocytes. Tyler and Dunn (1976: 73) believe that flounders fall into the second class and that winter flounders "sacrifice egg production and maintain body size so that when a good year comes the body will be large and be able to support a large ovary" rather than "sacrifice body weight and shunt substance into egg production so that some eggs are always produced."

Hislop, Robb, and Gauld (1978) studied the effects of nutrition on North Sea haddock fecundity. Their samples were not large enough to determine whether nutrition had an effect on female spawning, but they did find that well-fed fish produced *larger* eggs than did fish that were poorly fed (in contrast to the work of Bagenal, (1969) and Lyagina (1975). Hislop and his colleagues observed that high rations served to increase somatic growth but not fecundity, whereas low rations increased neither fecundity nor somatic tissue; rather, energetic resources were allocated to both reproductive and somatic tissue, but in a reduced amount. "A balanced energy budget of this sort seems appropriate to a species such as the haddock, in which an individual adult female generally survives for several spawning seasons. It may be the case that in short-lived species, when the adults only spawn once before death, the survival value of putting as much energy as possible into egg production would be much greater" (1978: 97).

In sum, it appears that the magnitude and "quality" of egg production are influenced by events occurring during growth in both the immature *and* mature phases of a fish's life. In the immature phase, growth is referred to somatic tissue, and its magnitude and patterns are relatively simple and predictable. When the fish becomes mature, growth is referred to somatic *and* reproductive tissue. These are related in that reproductive growth is mediated by the magnitude of the somatic biomass. The bulk of the somatic biomass, however, is created before the fish is mature and involves a complex of factors subsequent to maturity.

Mortality

Mortality rate is an important factor in the elaboration of cohort biomass. If, for example, a change in cohort biomass is observed, then without any additional information it would be difficult to ascribe the change (given constant recruitment) to either growth or mortality.

Whereas Chapter 5 emphasized the mathematical derivation of the prerecruit fish-mortality function (the recruitment-stock relationship), here we are concerned with postrecruit fish mortality. Postrecruit mortality *seems* to be a simpler process because its variability, in both a deterministic and a stochastic sense, appears to be much lower than at the prerecruit or early-immature stages. There is some evidence that postrecruit fish have a roughly constant or increasing natural mortality rate as a function of age. The increasing mortality rates are based on evidence from catch curves. Ricker (1975: 49), for example, presents convex catch curves for Queen Charlotte Islands (British Columbia) herring, which show an increased mortality rate with age.

Several agents cause natural mortality, including predation, senescence, environmental factors, and disease. Of these, only predation is a unique cause of death. The others can be direct causes of death, but they can also be contributing causes. For example, a fish might die of malnutrition; but if a fish is malnourished, it may be more susceptible to predation.

In addition to multiple causes of natural mortality, other evident complications occur. The most important one involves the relation of natural mortality to size. It is reasonable to assume that smaller fish are more susceptible to predation death than larger fish. If this is true, fish that never attain a large size, such as the smaller clupeoids, are subject to relatively high predation mortality throughout their life, and fish that attain a relatively large size, such as cod or tuna, must be exposed to minimal risk of predation death at large sizes. Hence the form of the mortality function differs among species of fish. Clearly any increase in predation would, in the simplest case, cause a decrease in the average age of the population. If predation were sufficiently high, it is conceivable that the average age of the population could be reduced below the average age of maturity, implying that the average fish would not live long enough to spawn. Perhaps as fish become large, senescence is not the result of old age but of malnutrition caused by difficulty in acquiring a large enough ration for maintenance, which explains the cannibalism adaptation in some fishes.

It is not sufficient to know simply the mortality rates or their causes; we must also understand the size-specific intensity of mortality and the nature of its possible density-dependent operation. For example, premature mortality will affect the spawning biomass per recruit only in an indirect way. If premature mortality reduces the population of juvenile fish to the extent that their growth is increased and they reach maturity earlier or at a larger size, then the spawning biomass will, all other things being equal, be larger. Senescent mortality may not have much effect on the spawning biomass because it seems to occur after the biomass of a cohort is already depleted by mortality.

Superabundance of Eggs

Fish are the only vertebrates that produce a superabundance of eggs. The term *superabundant* derives from the observation that, of the multitude of eggs spawned by a fish population, only relative few need survive to maintain a stable population. If there are so many "excess" eggs, then why should we be concerned about population dynamics, overfishing, or pollution? Why has nature, in its perceived parsimony, chosen to produce far more eggs than required to maintain a stable population?

The simplistic answer is that the superabundance of eggs, or the high fecundity of marine fish, is related to evolutionary processes. The more difficult question involves the mechanisms, the processes of evolution and natural selection, which are topics beyond the scope of this book.

It might be helpful, however, to sketch the nature of the process, particularly inasmuch as some researchers (for example, Ware, 1980) have taken high fecundity in fish as a measure of "fitness" (but see Svardson, 1949).

It is reasonable to think that most populations are statistically stable; that is, they have existed and will exist for a long time. In the context of egg production, the stable-population mean abundance may be denoted by

$$\bar{A} = NP, \tag{6.1}$$

where N is the annual population fecundity and P is the probability that an egg will survive to an age of actual or mean replacement. Hence to maintain $\bar{A}$ at some fixed level, an increase in N will require a decrease in P, and vice versa. A change in P thus reflects changes in the egg- and larval-mortality structure, and a change in N reflects changes in biomass or fecundity. Variability in N tends to be driven by events associated with adult fish; variability in P, by events associated with eggs or larvae. Thus some independence between N and P is probable.

Svardson (1949: 116) examined the tradeoff between N and P: "in fish a strong selection pressure has a tendency of increasing the egg number for each generation. As this obviously does not occur, there must be another selection pressure, counterbalancing the increasing tendency. From a theoretical point of view it thus is rather easy to conclude that there must also be a selection pressure for *decreasing* egg numbers, but it is not so extremely evident how this selection works."

Svardson thought that egg size was the most likely constraint on N. That is, the female would have a limited physical capacity to produce reproductive tissue, and this constraint would limit the total number of eggs produced by each female. Hence any increase in egg number would result in smaller eggs. Because the probability of survival of smaller eggs is lower than the probability of survival of larger eggs, the tendency to produce more eggs would be counterbalanced by the tendency for the smaller eggs to have a reduced survival rate (see, however, Elliott, 1984).

Perhaps the easiest way to think about the N-P process is in terms of sampling: the superproduction of eggs is a mechanism to sample the environment for salubrious living conditions for eggs and larvae. In other words, as Svardson pointed out, there is an advantage to a large number of eggs in that, all else being equal, the expected number of surviving eggs increases with number. Yet there are constraints on the production of eggs, not the least of which is the physical capacity of the female to produce a particular egg biomass. But these constraints have been ameliorated in various ways, as can be seen by rewriting the form of the sampling process identified in (6.1):

$$\bar{A} = NP = N_1 P_1 + N_2 P_2 + \ldots + N_m P_m, \tag{6.2}$$

where the subscripts refer to different qualities of eggs, eggs spawned at different times, or eggs spawned at different locations. Hence in (6.1) stability can be maintained only by increasing N for fixed P or by changing the "behavior" of the eggs or young fish to modify P. In (6.2) stability can also be maintained by the appropriate distribution of N's among P's, or vice versa. (It is important not to miss the analogy between [6.2] and various sampling strategies, such as stratified or cluster sampling; see, for example, Cochran, 1953.)

Equation (6.2) seems to represent a reasonable abstraction for the examples where the spawning process allocates eggs of the same species, sometimes with different properties, among time-space strata or clusters. For example, yellowfin tuna spawn over most of the vast tropical Pacific Ocean during most months. There must be some time-space partitioning of spawning, and it must have the property of stability, particularly as tunas seem to be among the least variable in terms of year-class strength in fishes. Northern anchovy have clearly partitioned their spawning process by means of the batch-spawn property. Whiting of the North Sea also spawn in batches, but spawning appears to occur more nearly over a continuum and over a much shorter time period than for northern anchovy. Herring of the northeast Atlantic behave much as the anchovy do, in that they spawn in nearly every month of the year. There are, however, major differences. Herring are partitioned into genetically discrete populations, each population spawning only once a year, but generally at different times. Differences in egg quality among the various herring populations are well known, as are differences among batches of whiting eggs (as in many fish, the eggs become smaller as the spawning season progresses).

Variability in the allocation of eggs to time-space intervals was noticed by Hjort (see Chapter 7) and formalized by Cushing (1967, 1969). As Cushing has shown for a number of species, the mean-mean date of spawning varies little, but the mean date can vary by a larger number, which indicates that the partitioning process can be quite effective. Obviously the temporal distribution of spawning is in effect a sampling of the environment. Variations in the timing of spawning can be linked to the production cycle or to seasonal variability in productivity. Cushing (1967) developed the match-mismatch hypothesis, which suggests that fish have evolved to spawn at about the time of a peak in the production cycle so that the larvae can feed on the increased production of small zooplankton and to some extent phytoplankton. He also postulated that fish had evolved to spawn on a fixed mean date, whereas the production cycle varied in timing. The largest year classes would then result in those years when the production cycle was centered, that is, when the timing of phytoplankton and, by inference, zooplankton production coincided with peak spawning. (Cushing termed this the match-mismatch hypothesis, referring to the years when spawning coincided with production as a "match" and to the years when it did not as a "mismatch.")

In his paper Cushing (1967) stated that the possible dependence of herring populations on production cycles was referred to by Iles (1964) in his study of herring maturation. In considering production cycles in the North Sea and environs, Cushing classified production according to data provided by Colebrook on phytoplankton color (Figure 6.12).

Basically the central North Sea, the Dogger and Buchan areas, and Iceland have a bimodal production cycle, with a spring and fall peak in the phytoplankton-color production index. The Dogger spring peak is mostly in April in the central and coastal regions and perhaps a month earlier in the eastern region. The fall peak is in October. The herring in the central North Sea spawn during the fall peak.

To the north in the Buchan region, the spring bloom tends to be rather later than that in the Dogger region, and the fall bloom is somewhat earlier. Again the Buchan stock spawns at the same period as the fall bloom. The Iceland stock spawns in September, but it is not clear that there is always a pronounced bloom in all regions in Iceland at about that time.

The production cycles seem quite different for the Downs, Plymouth, and Dunmore areas. There do not appear to be well-developed spring and fall blooms, and in these regions the herring spawn in the winter.

Farther north, in the Hebrides, the Norwegian area, and the northern North Sea, the bloom is predominantly a unimodal spring bloom, and the time of spawning is well correlated with the spring bloom.

Thus at the southern range of herring in the Dogger and Buchan areas, the production cycle is bimodal, and the herring spawn at the end of the second mode. In the northern part of the range, the production cycle is unimodal, and the herring spawn near the peak of the mode. The picture for the winter spawners is not clear, but there may be a mixture of stocks. These observations, then, set the stage for the match-mismatch hypothesis.

The hypothesis would be important if the match or mismatch were caused by variability in *either* spawning or production, but Cushing tends to think spawning is at a fixed time. He studied the "regularity" in time of spawning in the plaice, the cod, the Norwegian spring-spawning herring, the Fraser River sockeye salmon, and the California sardine and anchovy. The plaice, herring, cod, and sockeye exhibit small standard deviations and small standard errors in the mean date of spawning.

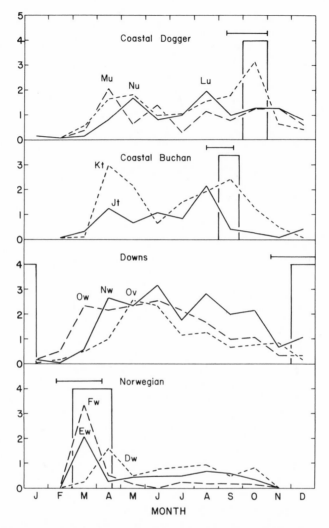

Figure 6.12 Selection of graphs showing the production cycle in various regions of the North Atlantic Ocean with spawning times of herring superimposed. (From Cushing, 1967: fig. 3.)

Although the mean date of spawning and its standard error are important statistics, the criticality of match and mismatch relates to the coincidence of spawning time with the appropriate components of the production cycle. In actuality, Runnström's (1941) data show what appear to be consistent differences in interannual spawning periodicity (Figure 6.13).

There are differences in timing among regions, and an apparently consistent trend toward later spawning from 1931 to 1936.

Cushing presented the mean dates of capture of arcto-Norwegian cod in the Lofoten fishery from 1894 to 1968. As he pointed out, the indices of central tendency tend to be surprisingly low. In addition, the ten-year running average increases from day 70 in the early 1900s to about day 82

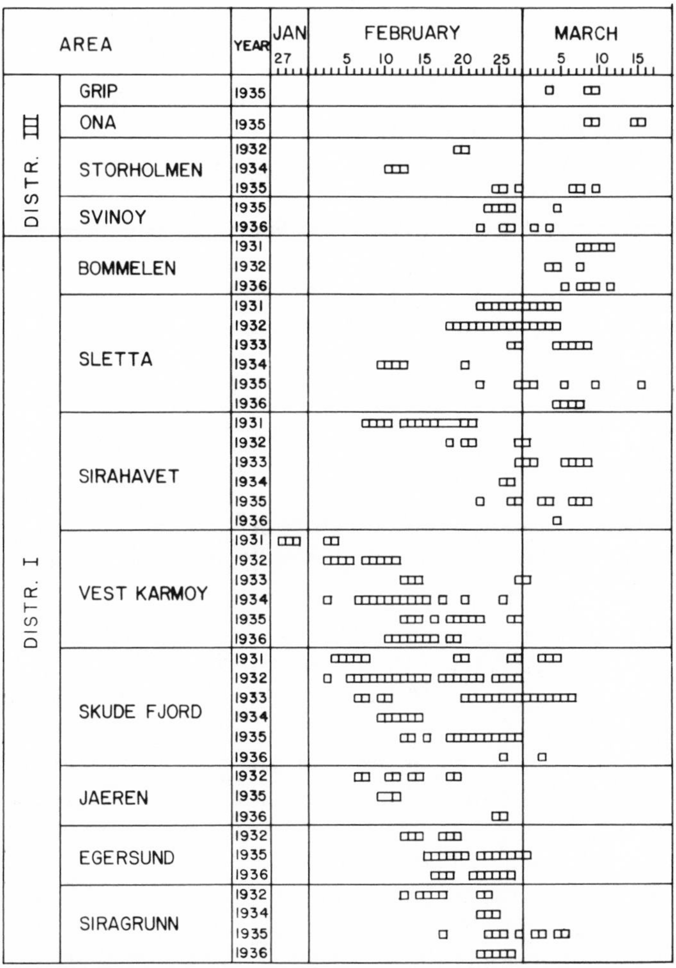

AREA		YEAR	JAN 27	FEBRUARY 5 10 15 20 25	MARCH 5 10 15
DISTR. II	GRIP	1935			
	ONA	1935			
	STORHOLMEN	1932			
		1934			
		1935			
	SVINOY	1935			
		1936			
DISTR. I	BOMMELEN	1931			
		1932			
		1936			
	SLETTA	1931			
		1932			
		1933			
		1934			
		1935			
		1936			
	SIRAHAVET	1931			
		1932			
		1933			
		1934			
		1935			
		1936			
	VEST KARMOY	1931			
		1932			
		1933			
		1934			
		1935			
		1936			
	SKUDE FJORD	1931			
		1932			
		1933			
		1934			
		1935			
		1936			
	JAEREN	1932			
		1935			
		1936			
	EGERSUND	1932			
		1935			
		1936			
	SIRAGRUNN	1932			
		1934			
		1935			
		1936			

Figure 6.13 Spawning seasons of herring off the Møre and southwest coast of Norway. Each square shows the occurrence of herring eggs in each region, year, and date. (After Runnström, 1941: fig. 8.)

in the 1930s, and then it declines to about day 77 in the early 1960s. There is, of course, more variability in the data than indicated by the averaged mean date of spawning. An examination of the data reflects that each point represents a mean with associated variability. If the estimate of standard deviation of the mean is about four days, then it is likely that a mean range of the observations would increase by about sixteen days or, in other words, the earliest date of spawning might be on day 59 in the 1890s and day 93 in the mid 1930s, a substantial change.

Pedersen (1984), however, presents data on the same fishery, that cover the period from 1929 to 1982. His analysis therefore excludes the beginning years of the time series presented by Cushing and extends the time series to 1982. Pedersen found that from 1930 to 1965 the MRI (the median of the ratio of roe weight in the catch to the weight of females in the catch) increased rather than decreased. He discusses the discrepancy between the two trends, which most likely relates to a changing relation between a median and a mean in an asymmetrical distribution in which the nature of the asymmetry is changing from year to year. Pedersen suggests that the mean age of the arctic cod declined from the 1930s to the 1970s from about eleven or twelve years to eight or nine years; because younger cod spawn late in the year, the spawning time is late. Unfortunately, he did not consider pre-1930s data.

Cushing presents another example of distribution of spawning times for California sardine and anchovy. In these fish, spawning is diffuse, and the central tendencies are not nearly as definitive as in the cod or herring.

Although the notion of central tendency is important, what is critical is the interannual match or mismatch with the productivity cycle. Therefore the mean date of spawning in terms of explaining interannual variability in fish production is mostly of interest for long-run considerations. Relative to the time spans of peak production (usually a month of so), the variability in time of spawning can be quite large.

Thus, in effect, Cushing has drawn attention to two components of spawning date variability. The first is what might be called standard-error variability and is quite small. The existence of the small standard-error variability ensures that those fish that have a fairly restricted spawning period (as opposed to fish that spawn several batches of eggs at the same time each year) spawn at the same mean-mean time each year. The second component might be called ordinary variability, and it is obviously of greater magnitude than the standard-error variability. It is important because it reflects the ability of fish not only to center their spawning activities at the most favorable time of the year, but also to extend these activities so that at least some of the spawning will coincide with the most favorable periods within the most favorable time of the year.

Sinclair and Tremblay (1984) doubt the generality of the match-mismatch hypothesis. They observe (1) that all herring larvae are constrained to reach 40 mm and metamorphose into the juvenile form between April and October; (2) that larvae from each of the herring populations are found in particular bounded areas; and (3) that growing conditions in each bounded area, and hence for each population, are different, so the time it will take to reach 40 mm varies. They conclude, then, that the date of spawning is the date at which the larvae would reach 40 mm less the number of days it would take a larvae to grow to that length. It does not seem that the assertions made by Sinclair and Tremblay contradict the match-mismatch notion.

Conclusion

The elaboration of reproductive biomass depends on the elaboration of cohort biomass, which depends on cohort growth and mortality rates. Stocks respond to a decrease in abundance by increasing their somatic

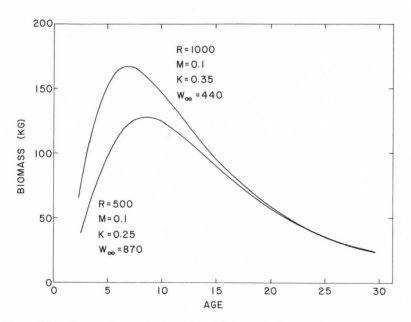

Figure 6.14 Biomass curves, in the spirit of de Veen (1976), showing the effects of increased growth rate on reduced recruitment in the sole.

and reproductive growth, and vice versa. The increase or decrease in growth is linked to the amount of available food. Thus the amount of food serves as a "signal" to the population on its own abundance. The food signal operates even during the juvenile stage, as density-dependent growth affects the age or size of maturity. It is also conceivable that the faster-growing juveniles are less susceptible to predation death, which would affect their mortality rate and the shape of the biomass curve, but no direct evidence has been found for this phenomenon. It is generally believed that, in contrast to juveniles, adults do not exhibit density-dependent growth; but this belief appears ill founded, in that adult fish respond to changes in density not so much in terms of somatic growth as in terms of reproductive growth.

The magnitude of changes in biomass may be seen in de Veen's study of sole, which shows density-dependent responses in growth. When fishing was at a relatively low level, the stocks were fairly abundant, and growth was limited. But when fishing mortality increased, the stocks were at a diminished level of abundance, and growth increased sharply. Figure 6.14 shows that growth compensation was so intense that the biomass curve for 500 recruits was approximate to the biomass curve of 1000 recruits. This finding, of course, suggests a major source of error in those recruitment-stock analyses that depend on biomass of the spawning stock, and not on reproductive biomass.

Beyond the simple elaboration of biomass, critical components of the egg-production process include the character of the eggs within each species or population and the manner in which the time-space distribution of spawning allocates eggs among various time-space strata in the environment. The theories concerning time of spawning are but fragments of this allocation problem, which merits further study.

7

The Life and Death of Fish Eggs and Larvae

Egg and larval mortality, and the phenomena that affect its magnitude, is a particularly complex subject, involving the eggs and larvae, the adequacy of larval nutrition, predators of eggs and larvae, and the microscale and fine-scale physical structure of the sea.

An organizational framework to understand this complexity can be developed from the axiom that larval mortality depends on the interaction of larval-food abundance and predation on larvae — that is, on the trophic position of eggs and larvae. Indeed, most of the components of the egg and larval mortality problem involve predator-prey transactions.

Historical Background

Many contemporary views on the causes of variability in recruitment and fish-larval mortality stem from the notions of Johan Hjort, director of sea fisheries in Bergen, Norway, from 1900 to 1916 (see Schlee, 1973). Hjort (1914) was concerned with the causes of year-class or recruitment variability in the "great fisheries of Northern Europe" (the Norwegian herring and cod fisheries, the herring and haddock fisheries of the North Sea, and possibly the North Sea cod fishery). He believed that fluctuations in recruitment "have their origin in certain conditions prevailing at a very early period in the life of the fish" (p. 104). This conclusion was evidently based, at least in part, on his observation that of the year classes of spring-spawning cod, haddock, and Norwegian herring, those in 1904 were exceptionally large. He reasoned that these year classes were produced by favorable hydrographical or biological conditions prevalent in the spring of 1904, and that the conditions were related to nutrition of the larvae and their passive movement or drift. (Hjort seems to have made a leap of faith, as he could have observed the strength of the year class only at the time of their recruitment. Thus all he could really assert was that the favorable conditions occurred between the time of spawning and the time of recruitment. This time interval was three years for herring and about one year for cod.) Discounting the possibility that the large year classes could have resulted from an unusually large number of spawning fish, he stated:

> The *very years in which the quantity of the roe at Lofoten was least [were] those which produced the richest year classes.* This being so it is difficult to avoid the conclusion that the actual quantity of eggs spawned is *not* a factor in itself sufficient to determine the numerical value of a year class. A rich spawning may produce a year class poor in numbers, while a large year class may have its origin in a year when the spawning was at its lowest.   (1914: 204)

Having concluded that egg production was not related to year-class strength (at least for the 1904 year class of cod), Hjort then directed his attention to the "nourishment of larvae." He referred to the work in France of Fabre-Domergue and Beatrix (1905) and Fabre-Domergue (1900), which suggested that "small larvae, even before their yolk is exhausted, commence to seek other nourishment, and those individuals which do not succeed in finding such become anemic, and die of hunger" (Hjort, 1914: 205).

These observations gave Hjort the idea that fish-larval nutrition is intimately associated with survival. When nutrition is "good," the survival rate is high, and a strong year class will result:

> At the time when the eggs of spring spawning fish abound, the water is almost destitute of all other organisms, animal and plant life . . . it will be observed [however] later on in the spring (at a time varying as to date in different years) that enormous quantities of microscopical plant organisms (diatoms, *flagellata, peridinea*) suddenly make their appearance being found in the form of a thick, slimy odoriferous layer on the silk of the net, which had previously been clean, containing nothing beyond fish eggs and some few crustaceans. It occurred to me, therefore, during these last investigations, that it should be well worth while to endeavor to ascertain how far the sudden appearance of this extensive growth might be of importance for the continued existence of the young fish larvae. If the time when the eggs of the fish are spawned, and the time of occurrence of this plant growth both be variable, it is hardly likely that both would always correspond in point of time and manner. It may well be imagined, for instance, that a certain — though possibly brief — lapse of time might occur between the period when the young larvae first require extraneous nourishment, and the period when such nourishment is first available. If so, it is highly probable that an enor-

mous mortality would result. It would then also be easy to understand that even the richest spawning might yield but a poor amount of fish, while poorest spawning, taking place at a time more favorable in respect of the future nourishment of the young larvae, might often produce the richest year classes. In this connection it must be remembered that one single cod may spawn millions of eggs . . . it might well seem possible that the high numerical value of these two year classes, especially that of 1904, should be due to the fact that the spawning set in so late as to ensure an adequate supply of nourishment for the young larvae at the stage when this was required. This question appeared to me of so vital importance, that I thought it best to lay the facts before one of the greatest authorities on microscopic plant life   (Hjort, 1914: 205)

Thus Hjort observed, at the dawn of fishery research, that (a) recruitment varied from year to year, (b) recruitment variability seemed not directly related to egg production or the magnitude of the parent stock, (c) larval nutrition influenced larval mortality, and (d) the timing of spawning relative to timing of the onset of spring production might be an important mechanism in the success or failure of a year class, an early observation of the match-mismatch idea later formalized by Cushing (1967). Hjort is also alleged to have coined the idea of a "critical period" —a period in the life of the larvae when mortality is especially high:

We must therefore look to the later stages of eggs to find the conditions which determine the numbers of individuals in any year class. This again leads us to the question, at which stage of the development the most critical period is to be sought. Nothing is known with certainty as to this; such data as are available, however, appear to indicate *the very earliest larval* and *young fry stages* as most important.   (Hjort, 1914: 204)

Hjort in fact may not have intended to imply that larval fish actually experienced a "critical period" (cf., however, May, 1974; the Fabre-Domergue and Beatrix references in May are different from those used by Hjort), as the phrase is not emphasized in his major 1914 work, and it is not mentioned in his 1926 work.

Nevertheless, the concept of a critical period seemed to have had some currency until it was examined critically by John Marr in his study of the "collapse" of the California sardine. Marr (1956: 169) placed modern focus on the problem and concluded that "what little evidence there is points toward survival at a constant rate or at a constantly increasing rate, rather than toward the existence of a "critical period."

While there was considerable controversy over the existence of a critical period, various experiments served to reinforce the concept. Blaxter and Hempel (1963), for example, estimated the amount of time after hatching that it would take herring larvae deprived of food to become too debilitated to feed and hence moribund. They defined this point rather picturesquely as the "point of no return" (PNR). They found that for herring larvae, the PNR depended on the weight of the egg and the ambient temperature. It appeared to be most strongly dependent on temperature: at $11° - 13°C$ the PNR was about forty days, which is not particularly dramatic, as herring in the sea are not likely to be exposed to an absence of food for such a long time. PNR studies were done with other species of fish larvae that seemed much more sensitive to starvation than herring larvae; they exhibited a PNR of about 3.5 days after hatching and 1.5 days after the onset of feeding. (Hunter, 1981: table 1, provides statistics on the PNR for various fish.)

Further accentuating the role of nutrition in the life and death of fish larvae, O'Connell and Raymond (1970) reported that under laboratory conditions high survival of northern anchovy larvae required concentrations of larval anchovy "food" of at least 1000 microcopepods per liter, a density far in excess in known concentrations in the sea. (As subsequently shown, microcopepods are not appropriate forage for first-feeding anchovy larvae; rather, the larvae appear to feed on dinoflagellates, and hence those that did survive in the O'Connell-Raymond study must have been unusual indeed.) The same conclusions emphasizing malnutrition as a primary cause of death might be drawn, varying only in degree, from a number of subsequent feeding studies (for example, Wyatt, 1972; Houde, 1978; Laurence, 1974; see Figure 7.1.)

Many field studies also pointed to the nutrition of larvae as an important mortality-inducing factor. LeBour (1920) found a high incidence of empty guts in fish larvae, and work by Arthur (1977), Beers and Stewart (1967, 1970, 1971), and Ellertsen et al. (1980) suggest larval-food concentration of perhaps $30 - 50$ nauplii and copepodites liter^{-1} concentrations considerably less than those reported necessary for survival. In 1978 Houde reviewed the literature and found only a few additional citations on nauplii and tintinnid densities that reflected higher *mean* densities, although some of the higher densities were from shallow-water systems that were possibly more fertile.

Actual observations in the field also implied that the availability of larval food was an important factor in the survival of fish larvae. Shelbourne (1957) found higher-condition factors in larval plaice that were collected in food patches than in those that were not. Hempel and Blaxter (1963) and Blaxter (1969a) found that condition factors of herring in the sea more closely approximated condition factors of malnourished larvae in the laboratory. O'Connell (1980), using histological techniques, found

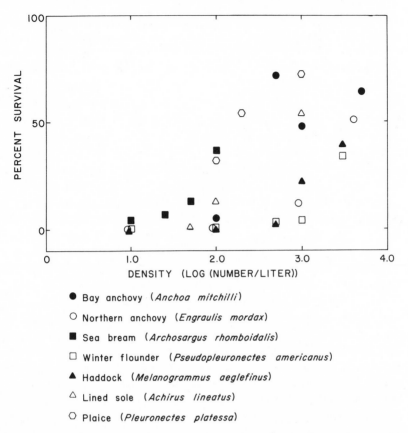

Figure 7.1 Relation of percentage survival of larvae to density of larval food in various feeding experiments. Note that in the critical region there are relatively few observations of expected survival rates and food densities. This region might be defined to include survival rates of less than 1 percent and food densities of less than 50 organisms per liter (1.7 organisms per liter on log scale). (Based on Thielacker and Dorsey, 1981: table 4; data points based on Wyatt, 1972; Laurence, 1974; Houde, 1978).

that nearly 10 percent of the anchovy larvae he examined showed signs of starvation-linked pathologies. His study has been interpreted as supporting the notion that the deaths of larvae result essentially from predation (see Blaxter and Hunter, 1982; Hunter, 1984: 537), but this conclusion does not appear to be correct. The nearly 10 percent of starving larvae is an instantaneous rate of roughly 0.1 per day, a typical mortality rate for fish larvae, which suggests that a substantial part of larval mortality is due *not* to predation but to either starvation or a combination of starvation and predation. Further observations of large numbers of moribund larvae have reinforced the impression that predation was not a major cause of death, although in some instances death was attributed to environmental "shock" rather than to malnutrition (see Strasburg, 1959; Marak, 1974; Soleim, 1942; Marr, 1956; Wiborg, 1976).

Thus up to the mid-1970s malnutrition continued to be thought of as a dominant factor in the life and death of fish larvae. Then experimentation took a new direction. Lasker (1975), in reviewing previous laboratory work, concluded that for northern anchovy (1) the size of food particles for first-feeding larvae is important; 3.5-mm larvae require food particles at least 50 µm in diameter (the reported range was 24-186 µm, with 70 percent of the items between 60 and 80 µm); (2) the density of food particles must be above a threshold density or larvae will not "feed successfully"; (3) only certain organisms are suitable for survival and growth (for example, *Gymnodinium splendens* in northern anchovy); and (4) attempts to feed by larvae are proportional to the density of food, and hence more food particles result in increased ingestion rates.

Taking account of these laboratory observations, Lasker (1975) conducted experiments on larval feeding at sea. He exposed laboratory-raised larvae to actual samples of water from a chlorophyll-maximum layer that had been evident in the coastal waters of California at the time of his experiments. He found that (a) for successful feeding, larvae require phytoplankton cells (actually *Gymnodinium*) in densities in excess of 20 cells/ml "at the same time or within 2.5 days after the larvae are ready to feed" (p. 460); (b) the cells must be approximately 40 µm in diameter; (c) not all phytoplankton cells in the chlorophyll-maximum layer are suitable for larval food (larvae feed on *Gymnodinium splendens* but not on spinous or chain-forming diatoms, such as *Chaetoceros* or *Thalassiosira*); and (d) the concentrations of microzooplankton from the study area (30 organisms per liter; 35 – 103 µm) were quite low relative to food requirement demonstrated in the laboratory. An important feature of these experiments was the ability to contrast feeding conditions during the presence and absence of the well-formed chlorophyll-maximum layer. When the layer was present, water samples from the layer reflected sufficient concentrations of cells to sustain larval feeding; but when the water column was mixed by a storm, the chlorophyll-maximum layer was obliterated, and densities of phytoplankton suitable for sustained larval feeding could not be found. Thus adequate feeding conditions could be temporary and linked to weather-related events.

Lasker's work continued to reinforce the notion of starvation as an important element in the life and death of fish larvae. At about the same time a series of models was developed that focused on this aspect of early larval life history (Jones, 1973; Jones and Hall, 1973, 1974; Vlymen, 1977; Beyer and Laurence, 1979). These studies, like many simulation models, are difficult to follow in detail; they give little consideration of the physical environment, and the stochastic nature of the larval-mortality problem is treated in varying ways with varying degrees of adeptness. On balance, however, each of the modeling efforts is well worth study, because each has a unique and different world view of the larval-mortality problem.

Jones (1973) was among the first to construct a model of the egg-larval regulation process (see also Cushing and Harris, 1973). Jones and Hall (1973, 1974) presented a further interpretation of the model, discounting the possibility that predation could affect the number of recruits. They observed that egg mortalities were about 30 percent per day and larval mortalities 5 – 10 percent per day. To account for these strikingly high but typical mortality rates, they suggested the possibility of "predation or dispersal of eggs into unfavorable regions" (1974: 88); sufficient predation on larvae to account for 5 – 10 percent daily mortality seemed unlikely. They argued that soon after hatching, larvae are dispersed in the upper fifty meters of the ocean and over thousands of square kilometers, which implies that the dilution of larvae is so great that predation cannot be effective in generating 5 – 10 percent mortality. Although the dilution argument does "not constitute proof that larval mortality rates of 5 – 10 percent per day cannot wholly be due to deliberate acts of predation . . . [it] does cast considerable doubt on this hypothesis and prompt us to enquire whether or not there might be another cause of high mortality rates" (ibid.). Jones and Hall thus suggested that the state of larval nutrition in the sea generates high mortality, although they did distinguish between mortality due primarily to predation, and mortality due secondarily to predation, that is, mortality that is enhanced by the diminished ability of malnourished or moribund larvae to escape predation.

Jones and Hall not only asserted that the lack of adequate nutrition contributes to the high mortality rate of larval fish, but they also suggested a mechanism that attributes death to malnutrition. They observed that the principal food of cod and haddock larvae are young stages of copepods. The size of copepods ingested is related to the size of the larvae, and gadoid larvae are not large enough to eat copepodite stage-V *Calanus* until the gadoids are 19 – 25 mm in length, a size attained after about two months. In the North Sea *Calanus* require two months to reach the copepodite-V stage, and "as a first approximation, therefore, it is suggested that successive cohorts of both species do indeed grow up together" (1974: 89). Jones and Hall estimated that a haddock larva at the commencement of feeding requires fifty Nauplius-I *Calanus* or seven Copepodite-I *Calanus* per day. "As it grows it eats larger copepod stages, and a 56-day-old haddock larva would require about 30 adult *Calanus* daily *to maintain* this growth rate" (p. 90). The larvae need to have about seven to thirty organisms per day. Hence successful juxtaposition of larvae and *Calanus* would result in high larval survival; unsuccessful juxtaposition would result in poor larval survival.

Jones and Hall (1974) stressed the role of food abundance and minimized the role of predation, which, if it occurs at all, simply acts to consume larvae that are moribund. These researchers began their analysis by asserting that larval food is patchy, and that the patchiness can be represented by the gamma probability distribution. (The gamma distribution represents the waiting time to the n-th occurrence. Only a special case of the gamma distribution is considered — the exponential distribution, which is generally taken to be the waiting time to the first occurrence.) According to Jones and Hall (1973: 37), the "probability of a larva encountering X_t organisms t days after feeding commenced" is

$$e^{-X_t/\rho_t s_t}, \tag{7.1}$$

where X_t is the minimum number of prey organisms contacted t days after feeding, ρ_t is the prey density, and s_t is the volume searched.

They then define X_t as a "minimum food barrier." In other words, if on any day a larva does not ingest X_t food organisms, it will die (this makes each t independent of any other t). Thus the number of larvae alive on day $t + 1$ is related to the number of larvae alive on day t by the probability that each larva finds at least X_t food organisms on day t; hence

$$N_{t+1} = N_t e^{-X_t/\rho_t s_t}. \tag{7.2}$$

They then relate the number of food organisms on successive days by

$$f_{t+1} = f_t - N_t U_t, \tag{7.3}$$

where U_t is a specified number of food items to account for the observed growth rate of the larvae. In (7.3) observe that the reduction in food depends only on feeding by the species or larvae under consideration, so an assumption of a strong density-dependent effect is built into the model, and the natural mortality of the prey (which must be of the same order of magnitude as that of the larvae) is ignored. It is therefore not surprising to find maximum survival at an intermediate level of food abundance and strong density-dependent effects, as suggested by Jones and Hall. Further interpretation of the Jones and Hall model is difficult because (1) organisms are usually not considered to be distributed as an exponential distribution; (2) the model is highly dependent on the abundance of food; and (3) it is unlikely that one day is independent from another.

Vlymen (1977) presented a detailed model of the life and death of a fish larva. His analysis considered the geometrical positions both of fish larvae and of larval food. Vlymen used the work of Hunter and Thomas (1974) as a point of departure. In studying the feeding behavior of northern anchovy larvae, Hunter and Thomas noted that the movements of the larvae could be represented by a random walk. That is, in the language of random walks, a larva at each instant of time can take a step forward, a step backward, a step to the right, or a step to the left. If a step in any direction is equiprobable, then the larva does not move in any particular direction; its motion is "random." If one direction is even slightly more probable than the others, the larva will tend to move in that direction. Hunter and Thomas (1974) observed that the directions of larval motion were forward when larvae were in low concentrations of food. When larvae were in dense patches of food, however, all directions of motion tended to be equiprobable, engaging the larvae in a random walk where they would stay more or less in place.

Vlymen parameterized the direction of motion and the variance of direction into a circular normal distribution. Using the theory of restricted covalent bonds, he then determined the mean distance traveled by a larva. He also developed expressions for attack rate and ingestion. Having specified the behavior of the larva, he directed attention to the distribution of the prey. He used notions that *might be related* to those associated with a negative binomial distribution to characterize prey distribution. Vlymen took into account the contagious distribution of prey by considering the concentration of prey in patches and between patches. The density of prey between patches was defined as C, and the density of prey within a patch was then $\overline{C} = C(1 + K^{-1})$, where K is a parameter referring to contagion. He assumed that the radius of a patch R^* is proportional to $\overline{C}^{\frac{1}{3}}$, and that the distance between patches is proportional to $(1 + K^{-1})^{\frac{1}{3}} R^*$. With this model Vlymen was able to show the relation of food concentration to patch size. While the model introduced the idea of patchiness in larval-feeding models, it too is difficult to interpret because the probabilistic assumptions underlying contagion are not well formed (see Beyer, 1982).

The Beyer-Laurence model also considers the nutrition of fish larvae, but it does not take into account predation or changing environmental conditions. It is important because it provides a stochastic representation of the feeding process and shows how the performance of larvae can change as the larvae increase in size. The model is basically a queueing-theory model. In other words, a larva is thought of as being fixed in space. The prey are in a queue, or line, in front of the larva and are continuously moving toward it. The times at which the prey reach the larva are called arrival times, and the time intervals between the arrivals of adjacent prey are called interarrival times.

All prey that reach the larva are not necessarily ingested. To be in-gested the prey must, in addition to being perceived, generate a feeding reaction and an "attack" from the larva. If the larva is already occupied in one of these activities or is in a latent period because of one of them, it obviously cannot begin to feed on a new prey. Thus it is possible for the larva to become satiated at high prey levels. Note that the satiation phenomenon is generally incompatible with the larval starvation syn-drome, thus calling into question the utility of predation models that posit satiationlike behavior for the study of fish-larval nutrition.

The essential feature of the Beyer-Laurence model is that it accounts for the ingestion of prey as a function of time and larval size. The larva grows as a function of the quantity of prey that it ingests. If prey are scarce and growth is reduced or negative, the larva eventually reaches the "death barrier," which is an exponential function of time:

$$W_t^* = W_0^* e^{\gamma t}, \tag{7.4}$$

where W_0^* is the minimum weight of the larva at time zero and γ is a constant. Hence survival requires that the weight of the larva at any time be greater than the "death-barrier" weight, that is, $W_t > W_t^*$.

To determine $W(t)$, one must study the ingestion of food. The quantity of food ingested is a function of the volume of water searched by the larva and the density of food items in the water. The volume of water searched is a product of the cross-sectional area of the larva's perceptive field, the larva's perceptive distance, and its velocity. The perceptive distance is a function of larval length, $r(\ell) = \delta\ell$; the cross-sectional area of the perceptive field is also a function of length $a(\ell) = \frac{2}{3}\pi r^2$, as is the larva's swimming velocity, $V(\ell) = v\ell$. Hence the volume of water searched by a larva is

$$s(\ell) = Va = \frac{2}{3}\pi\delta^2 v\ell^3. \tag{7.5}$$

The number of prey encountered in $s(\ell)$ is clearly a function of the density of prey in the volume of water or, in the context of this model, the density of prey waiting in line. But the amount of prey ingested by the larvae depends not only on the number of prey in the queue but on the way in which they are distributed in the queue. If, for example, the distance between each adjacent prey in the queue is a fixed constant, the interarrival times will be a fixed constant. If the prey density is low, the distance between them and the interarrival time will be large. If the time is sufficiently long, the larvae will be able to process each prey; but if the prey are sufficiently dense, there will not be enough time to process each prey. Hence these interarrival times are exponentially distributed with mean λ^{-1}, and the rate of prey arrival is

$$\lambda_t(\ell) = \rho(t)s(\ell), \tag{7.6}$$

where ρ is the density of food particles. After developing this fundamental equation relating the contact rate of the larva to its search velocity and the density of food, equations are developed to account for the reactive process; the attack process; the ingestion process; the rate of ingestion, food processing, and utilization; basal, routine, and active metabolism; and the costs of each attack.

These equations are used to develop a daily-growth equation:

$$\Delta W = (1 - \alpha)\beta(w)R\omega - Kw^n, \tag{7.7}$$

where α is the cost of food processing as a function of digestion, $\beta(w)$ is the fraction of food ingested that is digested, R is the daily ration, ω is the prey weight, K is the coefficient of daily metabolism, and n is a constant.

In addition to providing a calculus for the study of nutrition, the model is important because it shows substantial changes in the capabilities of larvae with increases in size. In particular, it shows the relation of growth to food density. Perhaps its biggest shortcoming is its extreme sensitivity to changes in the constant δ, a number with an operational meaning that is difficult to conceptualize.

Thus from a historical point of view, study of the linkage between recruitment and parent stock has been dominated by considerations of larval nutrition, with relatively little consideration given to (a) the fate of eggs, (b) predation on larvae, or (c) physical environmental factors that might, on appropriate scales, affect the fate of eggs and larvae. Models relating specific events are sometimes hard to interpret because their expansive simulation approach makes it difficult to extract all of the logic and consequences from the procedures. Despite criticisms that could be leveled at these approaches, these pioneering efforts have provided a foundation for relating biotic and physical interactions and recognizing the broader aspects of the problem, particularly the microscale and fine-scale ecosystem in which eggs and larvae live.

Conceptual Development

Serious consideration has been given only recently to the broader issues associated with the life and death of fish larvae. In addition to the simplistic idea that starvation is the major cause of larval death, more attention is being focused on (a) the fate of eggs, (b) the continuum of larval nutritional states, (c) the effect of predation on nutritionally adequate and malnourished larvae, and (d) the way in which the physical environment affects these properties of egg and larval dynamics.

These topics, many of which relate to the intensity of prey-predator interactions between larvae and their food and between larvae and their predators, are organized in Figure 7.2. The figure shows that an egg may either die as a result of various causes or it may hatch. A hatched larva may live long enough to become a recruit or it may die from (a) starvation, (b) predation that may be intensified if the larva is malnourished, or (c) predation independent of the larva's nutritional state. These possibilities, and the intensity under which they operate, are affected in varying degrees by the physical environment—notably, by temperature, motion, and light. Hence, as shown in Figure 7.2, each egg and larva may exist in twelve states. The probability that a fish reaches the recruitment state from the egg state is a function of the probability of the intermediary transitions. Figure 7.2 is thus a form of stochastic matrix that could be studied to gain insights into the egg-larval mortality system.

A detailed study of this matrix would be premature, because there is little empirical information on many of the states and transitions. The primary use of the matrix is to structure the complex of problems between spawning and recruitment into well-defined components. The twelve transitions, or subproblems, can be expressed more formally as

$$P[T_i|T, L, M, D] = Q_i, \qquad i = 1, \ldots, 12. \tag{7.8}$$

In other words, the ith subproblem or transition can be evaluated by determining the probability Q_i, given conditions on temperature, irradiance, motion, and density, T, L, M, and D, respectively.

The set of problems so identified forms a prey-predator system in which the organism of primary interest, the fish larva, is neither a predator nor a prey—it is both, preyed on by other species as well as by members of its own species, yet at the same time feeding on zooplankton or grazing on phytoplankton.

The prey-predator events associated with most of the transitions require a metric for measuring the intensity of the so-called forces of mortality on the fish in the setting postulated in Figure 7.2. An appropriate

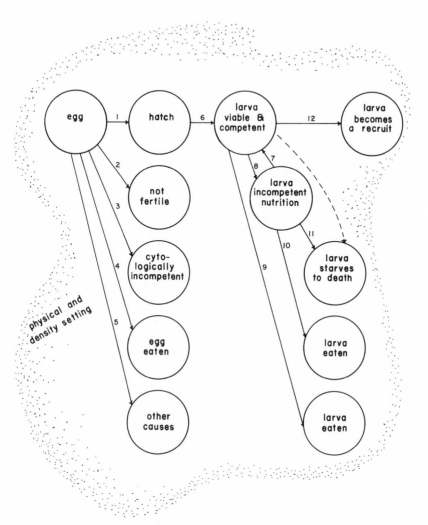

Figure 7.2 Transitions among various states linking recruitment and stock. The twelve transitions are set in an environment of varying temperature, light, and motion. The diagram partitions the problem into "study areas" and demonstrates its recent evolution from a narrowly defined "starvation problem" *(dashed line)*. There are, of course, other ways to partition the problem.

metric is the risk of death, which can be cast in terms of the familiar instantaneous rate model

$$N_r = N_0 \prod_{i,j} e^{-M_{ij}\tau_{ij}}, \qquad (7.9)$$

where N_r is the number of recruits, N_0 is the number of eggs, and M_{ij} is the instantaneous mortality rate from the ith risk of death operating at the jth intensity over time period τ_{ij}. To take account of the size or age-specific nature of the M_{ij}'s and the τ_{ij}'s, the expression would, of course, need to be more complex.

Thus the risk of death to which an egg or larva is exposed depends not only on the intensity M_{ij} but on the time period over which it operates. A larva exposed to a low value of M_{ij} for a particular i for a long period may have a risk of death equivalent to that of an egg or larva exposed to a high value of M_{ij} for a particular i for a short period.

Further development of the framework implied in Figure 7.2 requires linking the various risks of death with the time-space position of the egg or larva. The idea of an "occupancy volume" provides the geometry for such a linkage. The occupancy volume is a generalization of the Rosenthal-and-Hempel (1970) and Blaxter-and-Staines (1971) idea of a "search tube." The generalization seems necessary because it is almost certain that the intensity of search is not constant. Further, volumetric search by different fishes at different life-history stages must be conducted in various fractal dimensions, and estimates of volume searched would be difficult to compare without taking into account these dimensions.

As will be shown subsequently, the volume searched—that is, the number of prey contacted per unit volume—may depend to a considerable extent on the velocity of the prey as well as on the velocity of the predator. The central axis of the volume is traced by the trajectory of the self-propelled motion of the egg or larva (eggs are obviously not locomotory, but they do move relative to water because of buoyancy effects). The dimensional units of the central axis are in length per unit time. The diameter of the volume may be thought of in various ways: D_0 is the cross-sectional diameter of the larva; D_1 is the "diameter" of the cross-sectional area perceived by the larva; D_2 is the contact diameter, which is a function of not only the perceptive field but the velocity of the prey; and finally D_3 is the contact diameter as affected by the physical environment.

These definitions provide a geometry for different forms of occupancy volumes. The D_0 volume traces the larval path; the D_1 volume traces the volume of water perceived by the larvae; the D_2 volume indicates that the contact of prey by predator larvae is a function of the velocity not only of the predator but of the prey as well. In other words, the D_2 volume traces the "volume" in which particles are perceived. If the prey particles do not move, then the D_1 and the D_2 volumes are identical. The D_3 volume takes account of events external to prey and predator and their inherent velocities, such as the velocities and acceleration of the water mass in which prey and predator are embedded, as well as the ambient illumination and temperature.

The occupancy volume model thus provides a framework for discussing the transitions identified in (7.8) and for assessing the magnitude of the M_{ij}'s and τ_{ij}'s in (7.2). The occupancy-volume idea enables us to define the biological and physical environment intercepted by the volumes as they are advected through the water mass and the integration of environmental components as they affect contact rates.

Egg and Larval Life History

The linkage between what is known about egg and larval life history and the conceptual model just described is at a stage where it only serves to organize the issues of concern; it still requires considerable development as well as substantiation by field and laboratory studies.

Our discussion of egg and larval dynamics has a predominant fundamental concern: density dependence. The concern is not so much to ascertain whether density dependence exists, as it must for populations to maintain their stability, but to identify where in the population-dynamics process density dependence produces its effects. If, for example, portions of the process are density independent, it is relatively simple to consider the effects of environmental variation, because—all other things being equal—they will always induce the same reaction from the population. Yet the existence of density dependence at particular stages suggests that environmental effects cannot easily be separated from population-density effects.

Fate of Eggs

The density or apparent abundance of eggs in the sea diminishes through mortality, hatching, and dispersion. The initial number of eggs may be thought of more abstractly in terms of an initial number of occupancy volumes, which are affected by mortality, hatching, and dispersion.

Although many general observations on egg density exist, relatively few studies are sufficiently intense and detailed to permit estimation of

egg mortality. The densities reported are often difficult to compare because of variability in sampling techniques and reporting methods (for
examples, eggs meter^{-2} and eggs meter^{-3}).

Studies of the vertical distribution of eggs and larvae may be found, for
example, in Coombs, Pipe, and Mitchell (1981). They report on the vertical distribution of mackerel *(Scomber scombrus)* eggs and larvae in the
relatively deep water to the west of the British Isles and in the relatively
shallow water of the North Sea. Figure 7.3 shows that in March and April

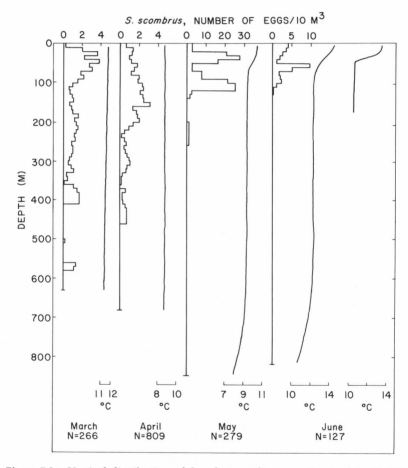

Figure 7.3a Vertical distribution of *Scomber scombrus* eggs (west of the British
Isles, 1977) and representative temperature profiles. The density of eggs is given
in terms of mean monthly abundance for all positive hauls. (From Coombs, Pipe,
and Mitchell, 1981: fig. 4. Reprinted with permission of ICES.)

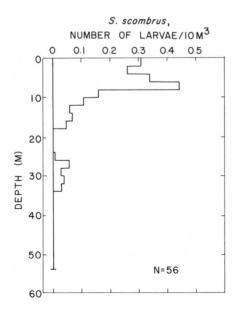

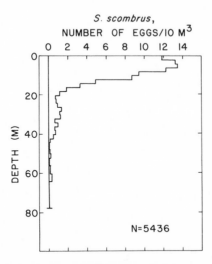

Figure 7.3b Vertical distribution of *Scomber scombrus* eggs (North Sea, 1978). The density of eggs is given in terms of mean abundance for all positive hauls. (From Coombs, Pipe, and Mitchell, 1981: fig. 5. Reprinted with permission of ICES.)

Figure 7.3c Vertical distribution of *Scomber scombrus* larvae (west of the British Isles, 1977) and representative temperature profiles. The density of larvae is given in terms of mean monthly abundance for all positive hauls. (From Coombs, Pipe, and Mitchell, 1981: fig. 6. Reprinted with permission of ICES.)

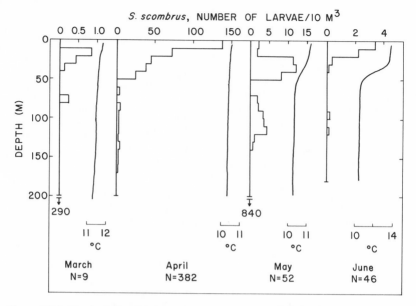

Figure 7.3d Vertical distribution of *Scomber scombrus* larvae (North Sea, 1977). The density is given in terms of mean abundance for all positive hauls. (After Coombs, Pipe, and Mitchell, 1981: fig. 7. Reprinted with permission of ICES.)

the water column is nearly isothermal and the eggs are distributed as deep as 500 or 600 m. Although the modal distribution tends to be in the upper 200 m, in May and June the water column begins to stratify and the eggs are found only in the upper 200 m. In contrast to mackerel eggs, mackerel larvae are, with a few exceptions, always in the upper 50 m of water. Mackerel eggs in the much shallower waters of the North Sea tend to be within 20 m of the surface.

Figure 7.4 shows the distribution of blue whiting *(Micromesistius poutassou)* eggs. These eggs are found mostly in deeper waters, often below 300 m, although there are considerable interannual fluctutions in the depth distribution of eggs, particularly in 1975. The distribution of blue whiting larvae shows an apparent ontogenetic trend, as the larger larvae are found closer to the surface.

Thus the horizontal and vertical distributions of eggs are subject to considerable variability, and it can be seen that sampling with other than discrete depth samplers could give a misleading impression of egg density. The mechanisms that produce the average distribution and the annual deviation from the average are not always clear. On the one hand, the mechanism might be ontogenetic, reflecting changes in the density or behavior, for example, of the eggs or larvae. On the other hand, the influence of physical properties of sea water, its motion and density, cannot be ruled out.

Clearly variation in these distributions could make predation on eggs and larvae and nutrition of larvae more or less effective, depending on the relative distribution of the predators and prey. For example, predation mortality could be most effective when the eggs or larvae are most concentrated and juxtaposed with the prey. Similarly, feeding by larvae could be most effective when the food items are most concentrated and juxtaposed with the larvae.

The exposure of organisms to physical variables is surprisingly complicated. Figure 7.3 shows that the variability of exposure of mackerel eggs to temperature is quite variable. In March, temperatures are between 11° and 12°C; in April they are between 10° and 11°C. In May, however, some eggs are at 11°C and some at 10°, whereas in June some

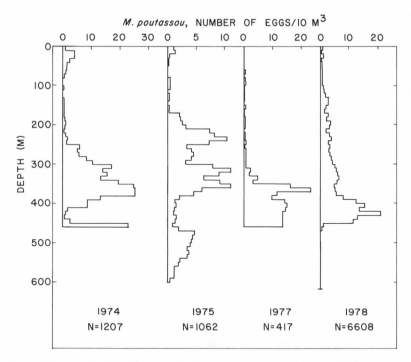

Figure 7.4a Vertical distribution of *Micromesistius poutassou* eggs. The density is given in terms of mean abundance for all positive hauls in each year. (After Coombs, Pipe, and Mitchell, 1981: fig. 2. Reprinted with permission of ICES.)

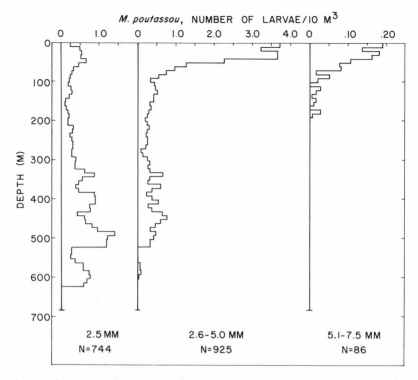

Figure 7.4b Vertical distribution of three size categories of *Micromesistius poutassou* larvae. The density is given in terms of mean abundance for all positive hauls in all years. (After Coombs, Pipe, and Mitchell, 1981: fig. 3. Reprinted with permission of ICES.)

eggs are exposed to temperatures of about 11 °C and others to temperatures of about 13° to 14°.

These differences might seem small, but they are not. Consider that the mortality of eggs might be 15 percent per day, which roughly approximates an instantaneous annual rate of 55. Thus the average egg lives for only several days; many live for only a short period. Further, the hatching time of eggs is quite sensitive to temperature (see Figure 7.5); hence very small changes in exposure could have a considerable effect on hatching time. The magnitude of very small percentage changes is not often appreciated. Suppose an imperceptible temperature change caused an imperceptible increase in survival rate — say, an increase from 0.05 percent to 0.052 percent. The increase in production from 10^{12} eggs would be 2×10^9 fish, quite a large number indeed.

Several researchers have reported high densities of eggs. For example, Graham (1924) reported densities of 200 cod eggs m⁻² in the Flamborough spawning area (North Sea). Smith (1973) summarized a massive quantity of data on sardine eggs; the maximum density was about 10,000 eggs under 3.3 m² of sea suface, a statistic considerably higher than the mean number of eggs. And Sette (1943) observed 900 eggs m⁻² for the Atlantic mackerel.

Fridgeirsson (1984) studied the vertical distribution of various species of fish eggs and larvae in Icelandic waters. A pump was used to sample different depths. The mean number of eggs m⁻³ for the three most abundant species was 39.5 for cod and haddock combined, 2.9 for dab, and 1.8 for Norway pout-whiting. The most abundant larvae were capelin, 3.85 m⁻³; cod, 0.455 m⁻³; and haddock, 0.078 m⁻³, which implied a crude egg survival rate of about 1 to 2 percent. As might be expected, 43 percent of the cod and haddock eggs were found at the surface; 61 percent, at 0 and 5 m. With respect to demersal eggs, densities for herring can be such that the eggs are found in layers several strata deep (see, for example, Blaxter, 1971; Hempel and Hempel, 1971; Taylor, 1971).

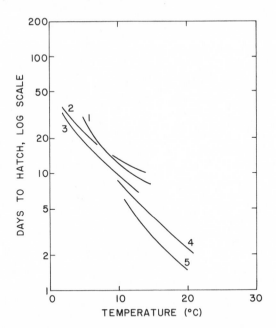

Figure 7.5 Relation of hatching time to temperature for (1) *Pleuronectes platessa,* (2) *Gadus macrocephalus,* (3) *Sardinops caerulea,* (4) *Clupea harengus,* and (5) *Scomber scombrus.* (Species selected from Blaxter, 1969: fig. 4.)

Typical reports of egg densities are much lower. Harding and Talbot (1973); Bannister, Harding, and Lockwood (1974); and Harding, Nichols, and Tungate (1978) report on extensive surveys of plaice eggs in the North Sea, where the densities of stage-I plaice eggs ranged up to, and on an areal basis seldom exceeded, 27 eggs m^{-2}. Talbot (1977, 1978), using data from the same program as Bannister et al. (1974) and Harding et al. (1978), reports various densities of stage-I plaice eggs, for which typically high average values appear to be about 1 egg m^{-3}. Houde and Lovdal (1984) sampled a single station in Biscayne Bay, Florida, for a two-year period, using obliquely towed (from the bottom) 35-μm and 33-μ mesh plankton nets. They found mean annual densities of 18 eggs m^{-3}. Anchovy *(Anchoa mitchilli)* eggs, with a mean annual density of 10 eggs m^{-3}, accounted for more than half the mean egg densities. (These mean annual statistics may not reflect the true magnitude of anchovy, because sampling on two days in 1967 and in 1977 yielded eggs densities between $\pm$ one standard deviation of 48 — 100, 35 – 60, 22 – 32, and 16 – 64 eggs m^{-3}.) O'Boyle et al. (1984: fig. 3) indicated that densities of cod, haddock, pollack, and silver hake on the Scotian Shelf were about 3000 eggs per 100,000 m^{3} of water strained or 0.03 eggs m^{-3} for each species.

In any event, the reports of mean density estimates for fish eggs and other plankton organisms may be of limited use, in that it is typical for the variance in plankton sampling to be considerably greater than the mean, which implies skewed or even multimodal distributions of plankton organisms. For such distributions, the mean may represent only part of the information required to appreciate the density of the population; a fuller appreciation would be gained by examination of higher moments, particularly the variance. This possibility is of particular concern in plankton work, because the densities of interest are often extreme, not average, densities. For example, it is often said that there is not enough food in the sea on average to feed fish larvae, a statement difficult to interpret, because the *average* fish larva dies, probably induced by malnutrition.

These observations on egg density provide a point of departure to examine the more critical problem of the temporal diminution of egg density. The diminution of egg density as a function of time, the *apparent mortality rate* of eggs, seems to be quite high (Hunter, 1982). Egg mortalities reported by Bannister, Harding, and Lockwood (1974) and Harding, Nichols, and Tungate (1978) are summarized in Figure 7.6. All these studies reflect egg mortality values that range from 0.02 to 0.14 per day.

The mechanisms associated with the high apparent mortality rate of eggs are not well understood. The mortalities might fall into density-de-

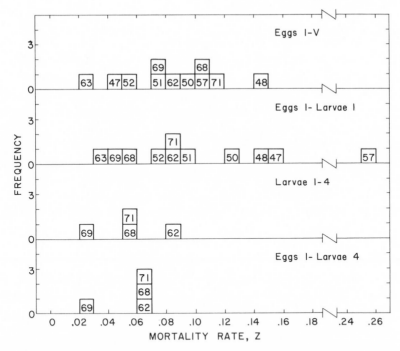

Figure 7.6 Frequency distributions of instantaneous daily mortality rates of plaice (*Pleuronectes platessa* L.) during various life-history stages. Year of observation pertaining to each datum is indicated. (Based on Harding, Nichols, and Tungate, 1978: table 48.)

pendent or density-independent patterns, and the specific causes, while again difficult to identify, can involve both predatory and nonpredatory death.

An example of a density-independent phenomenon may be found in Harding and Talbot (1973) and in Bannister, Harding, and Tungate (1974), who implicate physical environmental factors in the winter of 1963, one of the coldest on record in Europe, as affecting egg mortality. In 1963 the plaice, in which year-to-year recruitment is relatively constant, produced a very large year class, which was coupled with a very high survival of eggs.

In contrast, Daan (1981) presents data (Figure 7.7) for cod that suggest density-dependent effects on the survival of cod eggs. In Daan's figure high egg production results in a low-survival year class (except for 1970), and low egg production results in high survival, suggesting the operation

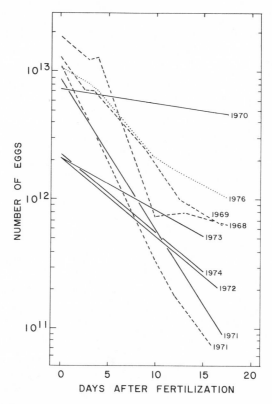

Figure 7.7 Cod egg production and numbers of eggs surviving up to time of hatching—Southern Bight. Note that, with the exception of 1970 and possibly 1976, egg-mortality rates are reduced at lower initial egg densities. The initial number of eggs is somewhat greater than 10^{12}, and the number of eggs eighteen days later is somewhat greater than 10^{11}; hence the survival rate is about 10^{-1}, or 10 percent. (From Daan, 1981: fig. 2. Reprinted with permission of ICES.)

of a density-dependent mechanism, with 1970 being an unusual year. Other examples may be found in fish that spawn demersal eggs (for example, Taylor, 1971, and Rothschild, 1961).

Evidence for nonpredatory egg death may be found in observations made by Southward and Demir (1974: 55), who report that pilchard (*Sardina pilchardus*) "egg samples collected off Plymouth and farther afield in the Western Channel have always contained a large number of dead or damaged eggs." The proportion of dead eggs in the samples ranged from 38 to 92 percent, with an average of about 50 percent. Southward and Demir cite observations of other researchers who also found a high abundance of moribund eggs in samples. Longwell and Hughes (1981), for example, found large numbers of dead mackerel eggs in the mid-Atlantic Bight off the east coast of North America.

The possible causes of nonpredatory egg death are not well understood, but they might be linked to cytological competence or to unfavorable environmental conditions. On the one hand, it is conceivable that the nutrition of the parents affects the viability of the male and female gametes, suggesting a mechanism dependent on the density of the parents. On the other hand, it is conceivable that microchanges in temperature might affect the motility of the male gametes either by affecting their physiology or by modifying the viscosity of the water in which the gametes swim.

In discussing a scant literature on "fertilizability," Blaxter (1969) observes that it is difficult to estimate the length of time of fertilizability. He cites Yamomoto (1961) as reporting a short fertilizability time for freshwater fish and a much longer time for seawater fish, and Nikolsky (1963) as reporting the short-lived fertilizability of *Oncorhynchus* and sturgeon. In contrast, herring gametes seem fertilizable for hours or even days.

Several environmental variables, including temperature, salinity, and oxygen, have been cited as causes of egg death (see Hempel's 1979 review). Two of particular interest relate to the propensity of pelagic eggs to be positively buoyant and hence near the surface. Very close to the surface the eggs are exposed to harmful ultraviolet radiation and mechanical shock from wind action, both possible sources of mortality. Anthropogenic factors have been linked to egg death, as well as these naturally occurring factors. Longwell and Hughes (1981) have implicated the high degree of pollution in the mid-Atlantic Bight for the large percentage of dead mackerel eggs found there.

With respect to predation, egg cannibalism by northern anchovy has already been discussed. Pommeranz (1981) and Harding, Nichols, and Tungate (1978) report large numbers of fish eggs and larvae in the stomachs of adult fish. Hunter and Kimbrell (1980) and MacCall (1980b) report that 20–28 percent of the natural mortality of northern anchovy eggs can be accounted for by cannibalism.

There is evidence that other species of fish and invertebrates prey on fish eggs. Hempel and Hempel (1971) describe predation of demersal herring egg by haddock and saithe. Evidence of invertebrates preying on fish eggs is somewhat limited, but in at least one striking example Hattori (1962) and Enomoto (1956) found that during short periods, in areas with intense blooms of the dinoflagellate *Noctiluca*, nearly 10 percent of collected anchovy eggs were found engulfed by *Noctiluca* cells.

Fate of Larvae

As implied in Figure 7.2, the fate of larvae is determined by their nutritional status and their susceptibility to predation, two factors that are unlikely to be independent. In general, the mortality rates of larvae seem to be lower than those of eggs. If known high mortality rates of eggs were sustained, then recruitment of virtually all year classes would be considerably reduced, although some data, such as those presented for the plaice, show no substantial difference between the mortality rates of eggs and of larvae.

It seems reasonable that density-dependent processes operate during both the prerecruit life of the fish and the larval stage. Yet, although there is some evidence of density dependence in the egg production and growth of juvenile fish, evidence of density dependence in the larval stage seems rather weak. This statement does not mean that density dependence does not exist. It may simply be difficult to detect, as the kinds of observations that would demonstrate its presence or absence in larval growth or mortality would require intense multiyear egg- and larval-density studies. (Contemplate the sample size required to detect differences in survival of say 0.050 and 0.052 percent.)

One of the difficulties in determining precisely the modes of density dependence is that the mechanisms may be operable only at extreme densities, and as a result a routine search through empirical data might mask or "average out" the density-dependent effects. As might be implied by Raitt's example on haddock (Figure 6.6), density-dependent effects may be evident only during years of unusually high or low abundance. Density dependence may be obvious only at extreme population levels, as these values would not be masked by sampling variability. The density-dependent processes might be effective only at extreme population levels, density dependence occurring only when density is either very high or very low.

The existence of density-dependent larval mortality is difficult to ascertain, but there are clues. Ware and Lambert (1985) report on the production of Atlantic mackerel *(Scomber scombrus)* eggs and larvae in the Gulf of St. Lawrence, giving the complete mortality rates for three year classes (1974, 1975, and 1976) (Figure 7.8). In 1973 and 1975 relatively large numbers of eggs were spawned. The mortality rate for eggs in the 1975 year class was higher than that for any of the other year classes; the initial number of larvae in each year class was the same, but high mortality rates seemed to persist in the 1975 year class. Taking these data at face value suggests that if density-dependent mortality operated, it was strik-

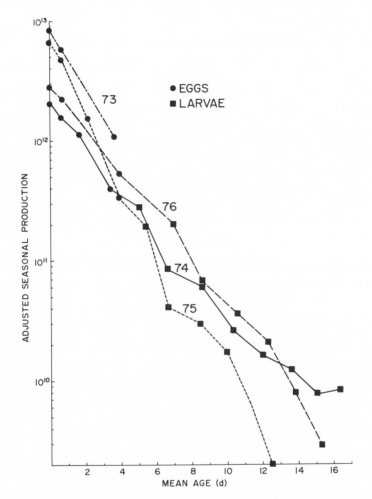

Figure 7.8 Decline in adjusted seasonal production of mackerel eggs and larvae as a function of age. (From Ware and Lambert, 1985: fig. 10.)

ing only during the egg stage of the 1975 year class. Another approach to estimating mortality was used by Methot (1983). He used the rings on sagittal otoliths to determine age and showed that the mortality rates in some months were substantially different from those in other months. Subsequent work would be needed to separate the physical-environmental and density-related effects on mortality.

MacCall (1979) presents a graph that shows the biomass of recruits as a function of larval density for the California sardine. Although MacCall fits a straight line through these data, a curved line would be an alternative interpretation, and the curvature would reflect a higher larval mortality (or juvenile mortality) at high larval densities. Burd and Holford (1971) present the same kind of data as MacCall, relating recruitment to larval abundance in herring (Figure 7.9). The herring data also show a proportionally decreased recruitment at high larval abundance.

The Ware and Lambert example gives a clearer picture of the operation of *larval* density dependence than the MacCall and Burd and Holford examples. Ware and Lambert measure the mortality of larvae as a function of their abundance, whereas MacCall and Burd and Holford measure the mortality of larvae confounded with the mortality of young fish.

The data presented by MacCall (1979), Burd and Holford (1971), and Raitt (1939), taken at face value, indeed suggest that density dependence is observable only when the population is at an extreme level of abundance. In other words, inferences on the existence of density dependence can be made only with sufficient observations at unusually high or low population levels (perhaps three out of fifteen years in the sardine and four out of fourteen years in the herring). Because these levels are relatively rare, there may be strong density-dependent controls that can only be suggested on theoretical principles rather than induced from empirical data.

Factors that might contribute to the magnitude of mortality and density-dependent mortality in larval fish include the nutritional status of the larvae, which can be ascertained by studying the nature and abundance of food in the sea and the nature of predation on larvae. Consideration of the nature and abundance of food involves examining the distribution of food items in the sea, the interaction of larvae with these food items, and the effect of the physical environment on this interaction.

Distribution of food in the sea. We think of the ocean as a volume in time-space containing myriad particles. In some regions the particles are widely dispersed; in other regions they are relatively close together. The particles may be distributed at random or they may be clumped into clusters or constellations. Some of the particles are alive and could represent organisms ranging in size from bacteria or viruses to large fish or mammals. Other particles are not alive and may consist of various forms of detritus or other suspended organic and inorganic sediments.

WIDTH OF THE FOOD SPECTRUM. Delimiting the set of particles suitable for consumption depends to a great degree on the size ranges of particles suitable for ingestion. Our problem is to partition the particles into two sets: particles that are relevant to the trophic requirements of the species of larval fish we are working with, and particles that are irrelevant. In

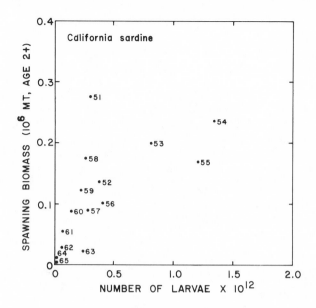

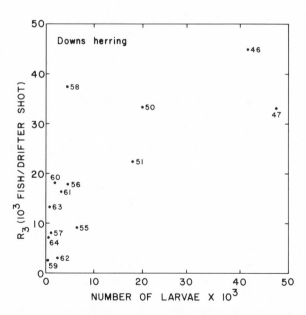

Figure 7.9 Two examples that suggest density-dependent mortality at stages between larvae and recruitment, based on reduced recruitment of California sardine and Downs herring at high larval densities. (After MacCall, 1979: fig. 4; Burd and Holford, 1971.)

addition to defining the set of interest, we must also specify some of its characteristics.

To begin this task we might consider the size range of particles appropriate as larval food, the nutritional value of each particle, and the density of particles. The size range of food ingested by larval fish appears to vary considerably. At the lower limit, fish seem able to ingest bacteria; at the upper limit, the size of food ingested appears to be a function of the width of the larval fish's mouth.

The observation that fish can ingest bacteria and small algal cells has considerable significance, as it had been thought that larval fish (at least larval northern anchovy, cf. Lasker, 1975) do not ingest particles less than 30 μm in effective diameter. This assumption seemed to be associated with an impression that copepod nauplii were the major food of larval fish. Ideas on the availability of food in the sea tended to be linked to the abundance of copepod nauplii. It now appears that fish larvae eat a wide variety of organisms, many of which could be smaller than 30 μm. This finding, of course, revises views regarding both the adequacy of food and the trophic dynamics of larvae.

At the lower edge of the size spectrum, Olafsen (1984) observed the ingestion of bacteria by cod larvae. The role of bacteria in the nutrition of fish larvae is not at present certain. It is not known whether bacteria can make a positive contribution to larval nutrition, either directly or by aiding the digestive process enzymatically.

With regard to phytoplankton, reports of green substance in larval guts are not at all unusual. The green substance could, of course, be derived from large phytoplankton cells (the dinoflagellate *Gymnodinium,* for example, is 50 μm in diameter). But there have been reports of fish larvae ingesting much smaller cells. Last (1978) found that fish larvae ingested diatoms; *Coscinodiscus* was found to be an important food in larval gadoids, particularly in cod *(Gadus morhua)* and in bib *(Trisopterus luscus).* Nordeng and Bratland (1971) found *Peridinium pellucidum* and *Coscinodiscus* sp. in the gut of cod larvae. Ellertsen and his co-workers (1980) fed cod larvae on *Dunaliella* (7–9 mm). The particular larvae had not yet formed movable mouths. The researchers imply that, despite the fact that the larvae had "green gut," the flagellates entered their mouths "by accident": "the cells clogged the visceral arches, and clusters of particles were swallowed. Larvae were able to spit out the particles when the jaw became functional" (p. 43).

Observations made by Moffatt (1981) reinforce the notion that small algal cells are important for first-feeding larvae of at least some species. Her observations were evidently stimulated by a comparison between papers by O'Connell and Raymond (1970), Lasker et al. (1970), Theilacker and McMaster (1971), Hunter (1972, 1976, 1980), and Hunter and Thomas (1974), which imply that the survival of fish larvae requires *much* higher densities of food items than those found in the sea, and papers by Detwyler and Houde (1970), Houde and Palko (1970), Saksena and Houde (1972), Houde (1972, 1973, 1974, 1975, 1977, 1978), Stepien (1976), and Houde and Schekter (1978), which imply that the survival of fish larvae requires densities of food of the same order of magnitude as those found in the sea.

A major difference between the set of studies that concluded that fish larvae require much higher concentrations of food than occur in the sea and the set of studies that concluded that the required densities are of the same order of magnitude as those in the sea is the fact the major food items in the former set consisted largely of the dinoflagellate *Gymnodinium splendens* and the rotifer *Brachionus plicatilis*. In the latter set, however, the experiments were conducted in a medium enriched by *Chlorella* cells, which have been thought to be too small (<10 μm) to be ingested by fish larvae.

Moffatt's results were surprising! She found larvae placed in a *Chlorella*-rich medium (>5000 cells ml^{-1}) had green guts reflecting ingestion of *Chlorella*. In experiments that included *Gymnodinium* or *Brachionus* and *Chorella*, there was a preferential ingestion of *Gymnodinium* and *Brachionus*, but *Chlorella* was also included.

The effects of *Chlorella* added to the culture medium were even more striking. Without *Chlorella* 4000 items of food per liter were required for 50 percent survival; with *Chlorella* only 100 food items were required. The explanation could lie in the improved nutritional value of the *Brachionus* as a result of their feeding on *Chlorella*.

The upper range in the size of food ingested by larvae appears to be a function of larval size, which, of course, is related to the dimensions of the mouth (Figure 7.10). In reviewing the apparent selectivity of larvae for different sizes of food, Hunter (1981: fig. 3) demonstrated that there were several patterns in the ontogenetic development of size selectivity. Very generally, the maximum size of food (width or length) increases as larval length increases. There are two forms of increase: in one the increase appears to be a smooth function of larval size; in the other it appears to be a step function of larval size. With regard to the lower range of food size, three patterns are exhibited: (a) the minimum size of food is roughly constant with increasing larval size; (b) the minimum size increases as a function of larval length; and (c) the minimum size increases as a step function.

Because of inherent variability and the fact that some measurements of prey have been made in terms of length metric, while others have been

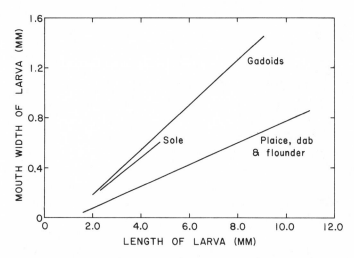

Figure 7.10 Mouth width of North Sea fish larvae as a function of larval length, showing the possibility of a physical segregation among food items. (After Last, 1978a: fig. 3; 1978b: fig. 3.)

made in terms of width metric, it is somewhat difficult to make precise categorizations. Nevertheless, the data presented by Hunter on the size range of ingested food suggest that as the sardine *(Sardinops sagax)* increases in size, it depends on a gradually changing food resource; that is, there is little overlap in the size range of food ingested by large and small larvae. In the anchovies *Engraulis mordax* and *E. ringens,* both the minimum and the maximum sizes of food ingested increase as a step function, indicating a well-known shift in feeding modality. In contrast to these two forms, the lower bound of food size ingested tends to be fairly constant, suggesting that larger larvae could eat the same size of food as smaller larvae.

The implications of these observations are most interesting. Does the sardine follow the same cohort of food as in grows? Do large jack mackerel or mackerel larvae compete for the same food resource as small jack mackerel or mackerel larvae, and what is the effect on density-dependent mortality and growth of either consuming the same cohort of prey or shifting to different cohorts of prey? And finally, what is the evolutionary significance of large anchovy larvae feeding on a different sized pool of forage from that of small anchovy larvae?

Even beginning to address these questions requires a detailed analysis of larval food and feeding behavior. One such analysis is provided by Last (1978a,b), who has studied various aspects of size and species selectivity in larval gadoids and flatfish in the North Sea. Figure 7.11 shows

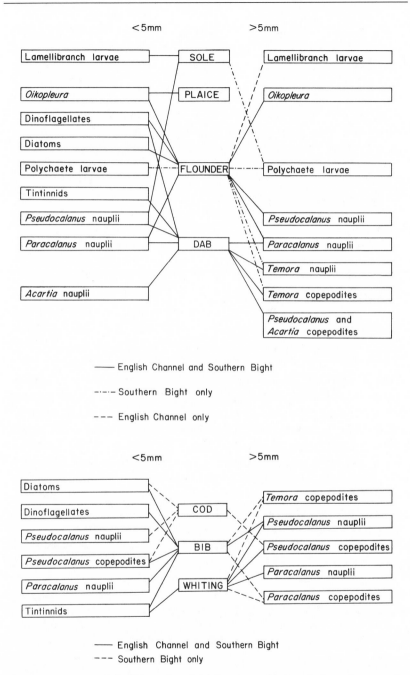

Figure 7.11 Diets of flatfish and gadoid larvae in the North Sea, showing the differences among flatfish and gadoid species as small and large larvae. (From Last, 1978a: fig. 1; 1978b: fig. 1.)

his analysis of diets of large and small gadoids and flatfish in the English Channel and the Southern Bight of the North Sea. The differences in diet are striking. For example, gadoids generally feed on fewer food items than flatfish. The small cod (<5 mm) relies on diatoms and *Pseudocalanus* nauplii and copepodites, but as it becomes larger (>5 mm), it is restricted to *Pseudocalanus* copepodites. The small bib ingests nearly all forms, with the exception of *Pseudocalanus* nauplii, but as it becomes larger, its diet is restricted to *Pseudocalanus* nauplii and copepodites. Like the cod, the diet of the small whiting is restricted, but rather than ingesting phytoplankton and *Pseudocalanus,* the small whiting ingests *Paracalanus* nauplii and copepodites and *Pseudocalanus* nauplii.

The impression that the diet of flatfish is catholic is based only on the flounder and dab. Both feed on a variety of forms when they are large and small. In contrast, sole and plaice diets are quite restricted, plaice feeding almost universally on *Oikopleura* and sole feeding as small individuals on lamellibranch larvae and *Pseudocalanus* nauplii and as large individuals on polychaete larvae.

Thus, in considering the effects of nutrition on the viability of fish larvae, each species seems to exhibit a particular pattern of feeding. To be sure, these empirical patterns are a function of the particular species of larva, its size and feeding behavior, and the size and behavior of the various forage organisms. It is nevertheless clear that there *are* patterns. An important ecological effect of these patterns is that species that are catholic in their feeding habits will be less dependent on dynamic changes in the variability of the species they eat, compared with species that have a restricted diet.

NUTRITION. Food particles can be classified not only by size and species but also according to their nutritional value, even though not much is known about the subject. In addition to the standard nutritional investigations, there is the question of the noncaloric value of food items. Considerable importance might be attached to vitamins or essential fatty acids, for example, but more research is needed. In addition, there is limited evidence that some fish will not eat certain types of food (Lasker, 1975). Other evidence suggests that larvae will ingest various types of food but will grow well only on some (Scura and Jerde, 1977, and Hunter, 1981). Certainly Moffatt's experiments and those of Olafson imply that there are important noncaloric aspects of trophic interrelationships that put into question various energy-flow analyses, these observations must be taken to imply that the nutritional "health" of a larva (or larger fish as well) depends only on the calorific value of the food, and not on the various organic substances that are known to be required for a salubrious diet.

Finally, an important aspect of larval nutrition might involve the nutritional value of the stomach or gut contents of the ingested prey, in addition to the nutritional value of the prey. It is conceivable that the gut of the ingested prey might contribute (a) necessary enzymes that the predator would not have been capable of producing, (b) material in some state of digestion that would not ordinarily be gathered or processed by the predator, and (c) essential nutrients, such as vitamins.

THE ABUNDANCE OF LARVAL FOOD. To add to the qualitative assessment of which particles in the sea are suitable as food for particular species and sizes of larvae, we can begin to think about the abundance of suitable particles. Usually we tend to think about the abundance of larval food in terms of its density. Although the density index is important, it is not sufficient to reflect that the abundance of food as the nature of predator contact with prey, from a predator point of view, depends not only on the density of the prey but also on how the prey are distributed in space and the velocity or accelerations of the prey relative to the predator. In other words, the question of prey abundance depends on the density, the statistical distribution, and the relative velocity of the prey.

Is it true that not enough food exists in the sea for the average larva to survive, and hence larvae that do survive are those able to find food or that are situated in unusually dense concentrations of food organisms? If this is true, we need to study the nature of the clusters and the implications that clustering might have for estimating the abundance of food.

The distribution of particles in the sea might be thought of as the distribution of points in space, a subject of considerable concern for ecologists, perhaps more for the purpose of characterizing various spatial patterns than to study the functions, the interactions, and the transitions among various spatial patterns (see Steele, 1978, for a discussion of plankton patchiness).

The distribution of points in space is usually analyzed in terms of a particular probability distribution (the Poisson distribution, say). Each probability distribution implies a particular set of "rules" for distributing the particles or points in space. If particles are distributed randomly, for example, then the particles should have a Poisson distribution.

Unfortunately, it is unlikely that nature consistently distributes particles in the sea according to any known distribution. Because of this assumption and the complexity of biological spatial distributions, ecologists have tended to address the more tractable problem of dealing with the moments of the distributions rather than with the distributions themselves.

The simplification has permitted a taxonomy of spatial distributions. If the mean and variance of a distribution are equal, then the distribution is said to be random, or Poisson; if the mean is greater than the variance, then the particles tend to be equally spaced and the distribution is said to be overdispersed; but if the mean is less than the variance, the distribution is said to be infradispersed, or contagious, the particles occurring in clumps or clusters (caution must be applied, however, in using the term *contagious*, since it usually refers to a distribution based on "after-effect" sampling, Feller, 1957).

While the taxonomical scheme facilitates dealing with a difficult problem, it can also be misleading. If we observe a particular relationship between successive moments in an estimated distribution that seems characteristic of a particular probability distribution, this does not mean that the observed spatial distribution of points was in fact generated by the distributional rules implied by the particular probability distribution. For example, if we observe that the estimated mean and variance of a spatial distribution are equal, it does not mean necessarily that the particles were distributed according to a Poisson or random process.

At least three interesting mechanisms can affect the relation of estimated moments. They also affect the distribution of roles implied by the observed distribution or, what is critical to our analysis, the apparent density of the prey as a function not only of its average density but of how the prey is distributed in space. First, if the prey tends to be contagiously distributed, then an increase in turbulence could reduce the contagion and generate quite possibly a random distribution. Second, random distributions seem to be more invariant with respect to scale than contagious distributions. That is, within reasonable limits a prey passing through a random distribution at various velocities will always perceive a spacing among organisms that conforms to a random distribution; however, a prey passing through a contagious distribution at the same velocities could perceive a spacing that conforms to either a contagious or a random distribution. Third, predation can change the nature of distribution by increasing or decreasing its contagion.

Because these points are so critical to the understanding of prey abundance in the context of the predator's frame of reference, it seems worthwhile to present a brief heuristic discussion of random and so-called contagious distributions (refer to Pielou, 1977, on which this discussion is partly based).

A simple model for considering a random distribution of animals in space involves the notion of a predator searching for a prey. The space in which the predator is committed to search is divided into n discrete cubes that may or may not harbor a single prey. The probability that any cube contains a prey is exactly p and the probability that it is empty is $1 - p = q$.

These assumptions establish the conditions under which it is possible to formulate the binomial probability law. This law enables us to determine the probability that the number of prey, a random variable X, takes on particular values x, $x = 1, 2, \ldots$, that is,

$$P(X = x) = \binom{n}{x} p^x q^{n-x}. \tag{7.10}$$

Hence the probability that the predator will contact $1, 2, \ldots$ organisms is a function of the number of cubes and the probability p. It is important to note that if the nature of the search process is affected by the predator's either finding or not finding a prey in a particular cube, then (7.10) is not a valid representation of the process; in other words, p must be constant throughout the search process.

The binomial probability law leads directly into the distribution most often used to describe the distribution of random points in space, the Poisson distribution. By letting $n \to \infty$ and p be proportional to n^{-1} with constant of proportionality λ, we generate the Poisson approximation to the binomial distribution,

$$P(X = x) = \frac{\lambda^x e^{-\lambda}}{x!}. \tag{7.11}$$

The Poisson property makes possible the study of the probability distribution of the mean distance between Poisson distributed points (see Pielou, 1977). The probability distribution, mean, and variance of the distance between points that are distributed according to a Poisson distribution can be deduced from (7.11). Consider a particular organism to be at the center of a circle. Let radius r of the circle be the distance to the nearest neighbor. There are no other organisms in the circle, which has area πr^2. The probability of having no organisms in a circle with area πr^2 is

$$P(X = 0) = e^{-\lambda \pi r^2}. \tag{7.12}$$

Hence the probability of the nearest neighbor being at least r units in distance is

$$P[R \geq r] = 1 - e^{-\lambda \pi r^2}. \tag{7.13}$$

which is, by definition, a probability distribution function. The density function is obtained from the derivative of the distribution function,

$$f(r) = 2\lambda \pi r e^{-\lambda \pi r^2} \tag{7.14}$$

from which the mean can be derived:

$$E(r) = 2\lambda\pi \int_0^\infty r^2 e^{-\lambda\pi r^2} dr = \tfrac{1}{2}\lambda^{-\frac{1}{2}}. \tag{7.15}$$

Similar computations can be made for the distribution of particles in three-dimensional space, yielding a mean distance between nearest neighbors of

$$\overline{D}_3 \cong 0.55\lambda^{-\frac{1}{3}}, \tag{7.16}$$

where λ is the number of organisms per unit volume.

With respect to contagious distributions, it appears that particles in the sea are only occasionally randomly distributed; they typically have a contagious distribution. Several empirical studies suggest that infradispersal or contagion is quite common. From this observation, however, it is difficult to appreciate mechanisms generating contagion. In fact, apparent infradispersal or contagion may result from the sampling process rather than reflect the actual distribution of the organism.

Contagious distributions can be considered to belong to a family of compound distributions, for which moments can be derived. A compound distribution is a distribution that is a distribution of a distribution. For example, the density of plankton patches may be a Poisson variable, and the density of organisms in the patch may also be a Poisson variable, but with a different mean. This particular distribution would be called a Poisson-Poisson or a Neyman type-A distribution. If, however, the distribution of patches was Poisson and the distribution of the number of organisms in a patch was logarithmic, then the resulting compound distribution would be negative binomial.

For example, if we sample the distribution of organisms in the sea and observe, as we would expect, a contagious distribution, then inferences can be made about the nature of the contagion. Pielou addressed this problem by considering the mean and variance of the number of clusters per unit (m_1 and v_1) and the mean and variance of the number of individuals per cluster (m_2 and v_2). If we write the probability-generating function associated with m_1 and v_1 as

$$G(z) = \sum_i P_i z^i, \tag{7.17}$$

and the probability-generating function of the number of individuals per cluster — probability-generating function the — associated with m_2 and v_2 as

$$g(z) = \sum_j \pi_j z^j, \tag{7.18}$$

then the probability-generating function of the compound distribution with mean M and variance V is

$$H(z) = G[g(z)] = \sum_i P_i \left[\sum_j \pi_j z^j \right]^i. \tag{7.19}$$

Taking first and second derivatives yields $H'(1)$ and $H''(1)$, from which it can be deduced that the mean and variance of any compound distribution are, respectively,

$$M = m_1 m_2, \qquad V = m_1 v_2 + m_2^2 v_1. \tag{7.20}$$

As is well known, the mean and variance of a Poisson distribution are equal, and so the mean and variance of the compound distribution, if we use (7.20), are

$$M = \lambda_1 \lambda_2, \qquad V = \lambda_1 \lambda_2 + \lambda_1 \lambda_2^2. \tag{7.21}$$

Now if we are willing to make the Poisson-Poisson assumption, then (7.21) can be solved for λ_1 and λ_2 to yield the mean and variance of the number of clusters or patches and the number of organisms in each cluster or patch from knowledge of the compound mean and variance:

$$\lambda_1 = \frac{M^2}{V - M}, \qquad \lambda_2 = \frac{V}{M} - 1. \tag{7.22}$$

Hence making the Poisson-Poisson assumption enables us to guess something about the underlying structure of the way that the particles are distributed. That is, we can appreciate how a fixed number of particles might be allocated among clusters or patches.

Equation 7.22 reemphasizes how misleading it can be to report only the mean density M in terms of understanding predator-prey interaction. A constant mean and a changing variance could produce considerably differing conditions for contact rates between prey and predator, as shown in Table 7.1. In this example, the mean of the compound distribution is constant, but the variance increases by orders of magnitude. Thus

Table 7.1 Hypothetical example showing implications of the variance on λ_1 and λ_2, even though the mean is constant.

M	V	λ_1	λ_2
10	11	100	0.1
10	100	100/90	9
10	1000	100/990	990/10

the change in variance alone is sufficient to indicate a considerable difference between the number of clusters per unit and the number of organisms per cluster. It also shows that the frequency with which a predator encounters a prey is a function of the variance in the distribution of the prey.

The interaction of larvae with their food and the effects of the physical environment. Now let us consider how D_0, D_1, D_2, and D_3 occupancy volumes might be embedded in the distribution of particles. The D_0 volume is characterized by the predator's, in this case the larva's, acceleration or velocity and its path along the central axis of the cylinder. In each larva's "personal" coordinate system, the central axis is rectilinear; however, the path of the central axis "over the ground" may be curvilinear in any place of the four-dimensional time-space coordinate system.

With respect to the larva's velocity along its path, considerable variability evidently exists and between species. Variability within species is linked, to a large extent, to the size of the fish and the possibility that the fish might alternate between "cruising" and "burst" swimming velocities, Hunter (1972) recorded the swimming velocities of a fast-swimming larva, the Pacific mackerel *(Scomber japonicus)*, and a slow-swimming larva, the northern anchovy *(Engraulis mordax)*, which seem to bound the swimming capabilities of larvae at various sizes (Figure 7.12). The figure suggests that a 1-cm larva might swim at velocities between about 1 and 2 cm sec^{-1}, and a 1.5-cm larva might swim at speeds between 2 and 6 cm sec^{-1}.

With respect to the larval path, it is evidently a function of the density and probability distribution of the food, as well as the behavior of the larva. Hunter and Thomas (1974) showed that larvae in a patch of food behaved differently vis-à-vis the direction of their movements from larvae that were not in a patch of food: the former turned more than the latter. These observations were used by Vlymen to study the behavior of larvae with respect to their contact with food at different interpatch and intrapatch food densities. Thus the shape of the patch may be an important part of the search process.

The movement of a larva among patches of organisms seems to be a special case of very complex problems regarding the extent to which a particle moving according to a random-walk process "covers" a patch of potential food organisms. Several scenarios can be considered in comparing straight-line motion with highly convoluted motion as typified by the random-walk process.

In straight-line motion in a field of randomly distributed particles, a larva will encounter particles at random, and it is of interest then to

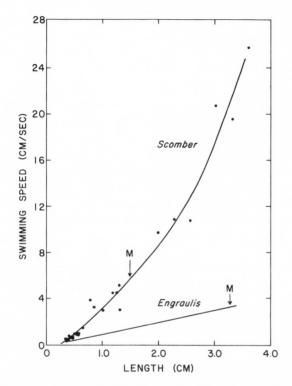

Figure 7.12 Swimming speed of Pacific mackerel *(Scomber japonicus)* larvae and juveniles recorded at 19°C. Points are means for five or more observations; curve is fit by eye; and swimming speed of northern anchovy larvae *(Engraulis mordax)* is recorded at 17°C–18°C. Speeds are total distance covered, including time spent at rest and in feeding, and M indicates fish length at metamorphosis. (From Hunter, 1972.)

determine the behavior of the larva with respect to attacking, capturing, and ingesting the particles relative to the Poisson distribution and identified with a particular density of prey. This special case is the model proposed by Beyer and Laurence (1980).

As an alternative, it is possible to consider a larva constrained to straight-line motion but moving relative to a field of prey that is contagiously distributed. Without respect to the particular type of contagious distribution, a larva would encounter alternately relatively high and relatively low densities of food, the particular densities being governed by the type of probability distribution and its parameters. Conceivably there may be some "optimal" distribution in which the spacing within and between particles optimizes larval growth or survival.

If the larva does not move in a straight line, then there are infinite possibilities regarding its path. The simplest is the random-walk notion. Before discussing this model, though, it is important to visualize the patterns of typical random walks, because the path taken by a particle subject to a random-walk process may be counter to what would be intuitively expected. A typical random walk does not uniformly cover the plane on which the walk is undertaken. A fish larva whose path is defined mathematically as a random walk covers a cluster on the plane and then goes to another cluster, and so on. In other words, the path of the fish larva is itself in a sense contagious (see Figure 7.13).

Now a larva moving in a random walk though a field of randomly distributed forage items will, on the axis of its normalized occupancy cylinder, encounter food items at random, making the process difficult to distinguish from the straight-line, random-particle process. This would require that the larva encounter its prey with replacement. That is, the concentration of prey in the vicinity of the larva is sufficiently dense that the larva's ingestion of food particles does not affect the density of the food particles. This situation is arguably unlikely if the larva is moving in a random walk as opposed to a straight line. In this case, then, the larva can affect either the density of its prey or its spatial probability distribution.

If the larva moves in a random walk through a field of contagiously distributed organisms, the possibilities regarding the probability distribution of contact rates are infinite. Some of the possibilities may reflect optimal cases for larval fitness.

The D_1 volume is analogous to the Rosenthal-and-Hempel or Blaxter-Staines search tube. The cross-sectional shape of the D_1 volume was thought to be a circle by Hunter (1972) and a semicircle by Rosenthal and Hempel (1970). At any rate, the Beyer and Laurence model shows that the volume searched will be altered significantly by slight changes in assumptions regarding the cross-sectional shape of the cylinder and its dimensions, as well as by factors that affect visual perception, such as ambient light, a function of contemporaneous irradiance, depth, and so on, and by the size of the prey.

The D_2 volume is different from the D_1 volume, as it has no precisely defined shape or linear measure. Rather, it is defined in terms of the prey that would be contacted by the fish larva as it proceeds through D_0 time-space. Thus the D_2 volume may be conical, with the dimensions of the cone governed by the relative velocities of the predator and prey.

As an introduction to the study of relative motion and contact rates associated with the D_2 volume, consider Koopman's (1956) detailed

Figure 7.13 Random walk with 30,000 steps. Note that the random walker alternates between a clustered pattern and straight-line flight. (See also Berg, 1983, and Mandelbrot, 1977.)

study of the relative motion of two objects in two-dimensional space. His study was made in the context of search theory regarding the interaction between surface vessels and submarines. The essence of his approach was to examine contact rates referred to the "searcher's", or predator's, coordinate system rather than to a geographical coordinate system. The use of the predator's coordinate system made possible the definition of a "region of approach" and the interpretation of various deterministic and stochastic models of prey-predator contact possibilities. Koopman's theoretical development, which was confined to two-dimensional space, was extended by Gerritsen and Strickler (1976) to three dimensions.

The region of approach is defined as the locus of points within which a prey can contact the predator as the prey moves through the predator's coordinate system. Figure 7.14 gives examples of the construction of the

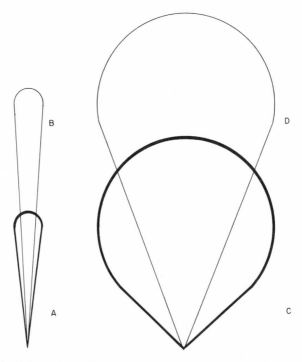

Figure 7.14 Four regions of approach for northern anchovy larvae and dinoflagellates. The swimming velocity of the anchovy larvae is either 2 cm sec^{-1} *(dark line)* or 4 cm sec^{-1} *(light line)*. In regions of approach A and B, the dinoflagellates have a velocity of 1.5 cm sec^{-1}. In regions of approach C and D, the dinoflagellates have a velocity of 1.25 cm sec^{-1}.

region of approach under conditions in which the velocity of the predator exceeds the velocity of the prey. It can be seen that the region of approach — that is, the D_2 volume — is as much affected by the prey's velocity as by the predator's velocity.

We now have in hand the material to address the problem of understanding the interaction between the oceanic environment and the variability in fish stocks, at least in terms of eggs and larvae. The problem of the interaction of the climate-ocean system and the variability in fish stocks is generally approached by linking events at distant ends of the causal chain, that is, by linking the atmospheric temperature or pressure field, for example, to recruitment strength without taking into account intermediary phenomena. Cushing (1982) has made a detailed and comprehensive review of this subject, and we can see from his studies that there is only occasional evidence for a relation of fish stock to climate, usually involving episodes of climatic warming or cooling. More recently Shepherd, Pope, and Cousens (1984: 266) reviewed the linkages between climate and fish stocks and concluded that "there is strong evidence, from the long history of fluctuations of abundance and the existence of fairly well defined geographical ranges, that an important relationship must exist between recruitment and climate. Determining the nature and the magnitude of the relationship, and elucidating the mechanisms involved, is however a most difficult problem."

It seems that the problem is at a stage where not much more can be accomplished by analysis and reananlysis of data. We know a priori that we can find environmental correlations rather easily, but it is extremely difficult to associate them with a meaningful measure of confidence, and even if we could, the great dimensionality of the system could quickly change conditions to invalidate relationships that might have been meaningful.

An alternative approach is to start at a disaggregate level of organization and examine the components of the problem to determine how they should be reconstituted or integrated. While such an approach might not yield specific parameter estimates, it might at least place needed focus on identifying the boundary between what can be known and what cannot be known.

The D_3 occupancy volume provides a basis for such disaggregation. The D_3 volume is a D_2 volume as it is affected by the physical oceanic environment. We might consider what is known about the physics of the sea that affects the D_3 volumes. To begin, we might think of the simplest classification of environmental variables: temperature, motion, and irradiance. These variables are, of course, not independent; for example, temperature affects the density of sea water, which affects the motion of sea water. Irradiance is often correlated with temperature, as both are functions of depth.

Despite these complications, we have a general idea about the effects of temperature, irradiance, and motion on the geometry of the D_3 volumes. We know that rising temperature increases (or decreases) the rate of many physiological or behavioral activities. For example, temperature influences swimming speed, hunger, and growth in fish larvae and the development of eggs. The temperature effect alone would affect the length of the central axis of the D_3 volume and could increase or decrease the consumption of prey, given the occurrence of contact. Microscale and fine-scale temperature differences are well known but not extensively

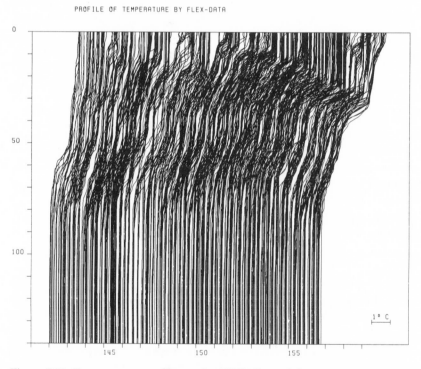

PROFILE OF TEMPERATURE BY FLEX-DATA

Figure 7.15 Temperature profiles at the FLEX Central Station versus time in Gregorian days. The profiles show rapid changes in temperature structure over a several-day period. It can be inferred that changes in temperature structure are related to changes in water motion. (From Friedrich, 1983: fig. 4; based on original computer plot kindly supplied by H. Friedrich.)

documented. Friedrich (1983), for example, shows rapid changes in temperature structure in the North Sea (Figure 7.15).

Variations in irradiance must also have an important effect on variations in the geometryof the D_3 volume. As is well known, fish larvae are visual feeders (Blaxter, 1968, has studied light thresholds above and below which herring larvae cannot feed). Surely feeding ability is not a step-function response to light—there must be a gradient. Local variability in the temporal-spatial distribution or irradiance in the sea relates to variations in turbidity, cloud cover, and biomass. These variables, particularly the last, are subject to high-frequency variation (Figure 7.16), causing similar variation in the D_2 volume and hence in the geometry of the D_3 volume.

Motion is perhaps the most subtle of the variables, in that the consequences of actual microscale and fine-scale turbulence and shears are rarely thought about with regard to biological variability (see, however, Wroblewski, 1984). It is conceivable that the turbulence and shears could modify the region of approach. In this regard, our discussion of regions of approach shows how differential velocities of prey and predator might transform the D_2 volume into a D_3 volume. Some interesting approaches have been suggested by Woods and Barkman (in press). They theorize that there is a diurnal turbocline above the seasonal thermocline. Water flow is turbulent above the turbocline and laminar below. The turbocline is close to the thermocline in the morning, rises to the surface by midafternoon, and dissipates in the evening. The existence of a vertically dynamic turbocline, along with theories on Langmuir circulation (see, for example, Ledbetter, 1979), suggests a rich, unseen structure that could have a considerable effect on D_3 occupancy volumes.

Perhaps the most dramatic examples of rapid change in the microscale and fine-scale environment are found in Lasker's observations on the effect of a "storm" on feeding conditions for the northern anchovy. Storm events are not simply independent changes in a single physical variable; rather, they constitute an integration of events—usually more wind, more mixing, more clouds, and possibly a new temperature structure. While Lasker (1975) has shown that at least some anchovy larvae find better feeding conditions in the phytoplankton-rich layer, these conditions are not sufficient for larval survival, as Husby and Nelson (1982) have implied. It may even be that while larvae *in* the phytoplankton-maximum layer have better feeding conditions, *all* larvae have better average feeding conditions when stability deteriorates, and in addition they might be less vulnerable to predators. The complexity of the stability argument is also seen in Laurence's (1982) graph of a storm event off

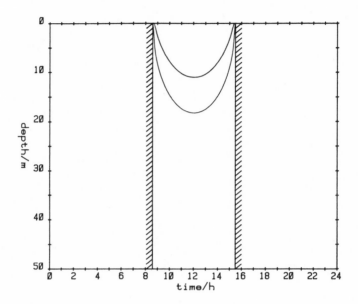

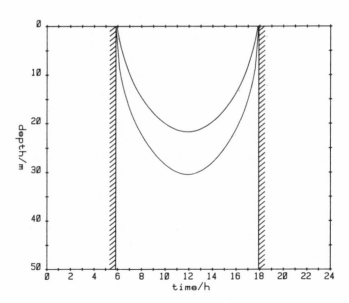

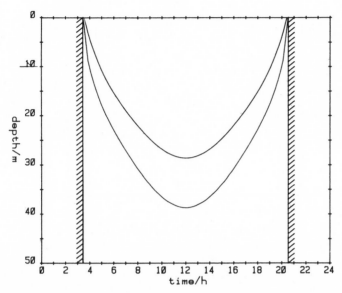

Figure 7.16   Depth of 1 W m⁻² isolume under conditions of clear sky *(lower curve)* and 100 percent cloudiness *(upper curve)* at N. lat. 55° during the winter solstice (22 December), the equinox (21 March), and the summer solstice (21 June). Extinction is appropriate to Jerlov (1976) type-III water, and attenuation of light by clouds follows Kasten (1981). The 1 W m⁻² isolume is the approximate threshold or limiting value of irradiance below which larvae cannot perceive their food. Figure shows how cloudiness can reduce food consumption, although the magnitude of the effect is not certain. Note also how the volume of illuminated water rapidly expands and contracts with the season, which reinforces the match-mismatch idea. (Figures prepared by A. Horch, Institut für Meereskunde, University of Kiel, courtesy J. D. Woods.)

New England: on one day larvae are subject to a certain temperature regime; then a storm deteriorates thermal stratification, thereby generating an entirely different thermal regime.

Thus one can consider a wide variety of physical factors in connection with D_3 occupancy volumes. What is of considerable interest beyond the deformation of D_2 volumes by the oceanic environment to create D_3 volumes is the way in which D_3 volumes interact in time and space. It is, for example, conceivable that at low densities of organisms — that is, low densities of D_3 volumes — intersection among D_3 volumes would be minimal, and density-dependent events would not occur. Now keep the density of D_3 volumes constant and increase temperature, irradiance, and small-scale turbulence; the D_3 volumes begin to intersect, creating the opportunity for density-dependent trophic transactions. Put perhaps too simplistically, seemingly minor, usually unmeasured changes in physical variables (particularly when scale is taken into account) could have a considerable effect on intra-annual trophic transactions in the sea.

The study of D_3 volumes, however, will require a different approach to sampling, as the very short lives of larval fish must be taken into account. An average larva might be only ten days old, so the temporal-spatial density of sampling would have to be quite intense.

Predation

Early studies on the mortality of fish larvae concentrated on the adequacy of nutrition as a primary factor. Recent thinking, however, has tended to increase the importance of predation. Hunter (1980, 1982, 1984), who has written several review papers on the subject, reports many observations identifying predators of fish eggs and larvae, suggesting that further studies will increase the list. Very little work has been done on the theory of predation, particularly in applying available models to the predation of fish (for example, Holling's 1965 model or the Lotka-Volterra model).

Clearly if the factors of larval mortality are to be better understood, the problem has to be placed in the context of Figure 7.2. We need to investigate the effects of predator fields on fish larvae, in terms of both a fixed nutritional status for the larvae and a variable nutritional status (see Bailey and Yen, 1983).

As far as a theoretical framework is concerned, the distinction between predation on larvae and larval nutrition may be a bit artificial, in that the predation and nutrition questions both involve trophic transactions. In fact, the D_0, D_1, D_2, and D_3 occupancy volumes provide a basis for the construction of an environmentally dependent theory on the linkage among trophic transactions. In addition to occupancy volumes for fish larvae, we now have occupancy volumes for predators of larvae; in fact, we have taken the evaluation of the probabilities specified in (7.8) a bit further. The evaluation of these probabilities will require the study of the interaction in time-space of two sets of occupancy volumes: D_{01}, D_{11}, D_{21}, and D_{31} with D_{02}, D_{12}, D_{22}, and D_{32}, where i refers to the particular kind of volume and j indicates whether the volume is specific to the predators of larvae ($j = 1$) or to the prey of larvae ($j = 2$).

Conclusion

Basic modeling efforts on the fate of eggs and larvae have concentrated on the nutrition of larvae; new efforts are required to consider the interaction of nutrition, predation, and the physical environment as influencing the mortality rates of very young fish.

The nature of density dependence is a critical element in this study. Density-dependent effects might have greater influence on the egg stage than on the larval stage, but a precise evaluation is difficult owing to what we might call the Raitt phenomenon, in which the existence of *observable* density dependence is itself a function of density. The possible lack of density dependence in larvae within, say, $\pm 2\sigma$ of the mean annual recruitment magnitude might be the one opportunity for the stock to exhibit positive feedback and to explore as many microenvironments as possible for salubrious living conditions, a supposition analogous to Andrewartha and Birch's (1984) discussion of den Boer's (1968) "spreading the risk" hypothesis. (This would, of course, imply a lack of density-dependent death or growth at the very earliest stages of larval life.)

It is important to think of the mechanisms that might induce density dependence. The argument that predation on eggs and larvae is more effective at higher egg and larval densities than at lower densities can be extended to include the food resources of the larvae; it could be reasoned that if the larvae were malnourished, they would be more susceptible to being eaten. Assertions that larvae are too dilute to affect each other's food suggest that models invoking density-dependent feeding responses are invalid. For example, Houde and Shekter (1980) studied the effects of food densities on the feeding behavior of fish larvae, using both Rashevsky's (1959) and Holling's (1965) models. The use of these models on feeding interactions in the sea is open to question because their primary feature is their portrayal of a saturation phenomenon. But if food is so dilute that larvae do not affect each other's food supply, one would have to conclude that saturation occurs mostly under artificial laboratory conditions. Cushing (1983) analyzed this matter and concluded that there is an opportunity for larvae to affect each other's food supply.

Another analysis of this issue can be generated from some of the calculus presented in this chapter. If organisms are randomly distributed, with an initial density of N_0 and a density at any subsequent time of $N_0 e^{-zt}$, then according to (7.16) the mean distance between nearest neighbors is

$$D = 0.55[N_0 e^{-zt}]^{-\frac{1}{3}}. \tag{7.23}$$

Figure 7.17 gives values of this function for one and three organisms per cubic meter, and instantaneous daily mortality rates of 0.05, 0.01, and 0.10. The mean distance between fish larvae increases with time and mortality rate. If the trend in Z declined with time, as some authors have suggested, then Z could be replaced with $Z(t)$. Figure 7.17 suggests that fish larvae can affect each other's food resources because the swimming speed of most fish larvae in a twelve-hour day, for example, far exceeds the mean-distance spacing between larvae.

The existence or extent of density dependence is a function, to some degree, of the mobility and feeding capabilities of fish larvae. The ontogenetic changes in capability are certainly important, as are the interspecific differences in capability, which suggests that different mechanisms are associated with density dependence in each species. For example, the functional development of various species of fish larvae at hatching may be quite different. According to Blaxter and Staines (1971), herring *(Clupea harengus)* and plaice *(Pleuronectes platessa)* are relatively well developed, as indexed by the fact that they have pigmented eyes; in contrast, the pilchard *(Sardinia pilchardus)* and the sole *(Solea solea)* have nonfunctional, unpigmented eyes. The impression that closely related fish, such

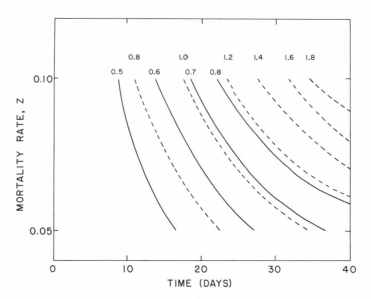

Figure 7.17 Isopleths of mean distance between Poisson-distributed organisms for specific daily mortality rates and days or exposure to the particular mortality rate. Initial density for the solid line was one organism per meter^3 and for the dashed line three organisms per meter^3.

as herring and pilchard or sole and plaice, can differ in early life history is reinforced by Moser (1981), who opines that size is the most prominent specialization in flatfish. Some species of flatfish have among the smallest eggs, while others have the largest eggs of marine fish, ranging from 0.6 mm in diameter to an astounding 4.0–4.5 mm for the Greenland halibut *(Rheinhardtius hippoglossoides)*. The range in egg size is matched by the range in larval length; the smallest larvae are 1.6 mm long; some are typically about 100 mm in size; and deep-sea forms have been observed to be 220 mm in length. The functional capability and sizes of various species of fish larvae are of critical ecological importance because they reflect the food requirements at first feeding. The smaller, less-developed larvae possibly find algal cells to be suitable food; the larger larvae tend to require organisms at least as large as copepod nauplii.

Thus the larva's ability to find and ingest food and to escape from or increase its susceptibility to predation depends on its stage of development at hatching. This stage can be a function of the species (for example, herring versus sole), of the population (herring from large-egg populations versus herring from small-egg populations), or of environmental conditions (larvae from large or small eggs, where egg size is a function either of maternal nutrition or of the average age of the spawning population).

To analyze the influence of density dependence on mortality, we must have a rather complete understanding of the causes of death. Figure 7.2 suggests that the factors associated with larval death—larval nutrition, predation on larvae, and the physical environment—are not easily separable. The notion of occupancy volumes helps us separate the various factors of mortality. We can begin to think not only of the distribution of constellations of particles in the sea but of constellations of occupancy volumes. It is easy to imagine (1) intraspecific effects associated with increased overlap of volumes as density increases; (2) interspecific effects associated with increased overlap of the volumes belonging to species A with the volumes belonging to species B; and (3) environmental effects that modify the geometry of volumes belonging to species A and to species B differentially as well as affect their intersections, thus possibly modifying the fitness of some populations relative to that of other populations.

The problem is now somewhat different, as we are no longer concerned with nutrition and predation, but rather with what seems a prerequisite to the study of larval nutrition and predation on larvae, the more general notion of intersections of occupancy volumes of fish larvae and of their prey and predators.

8

The Population-Dynamics Process

The preceding observations on recruitment variability and its causes lead to the concept that recruitment is stabilized by a population-dynamics process. The process involves signals that cause the population to increase when its abundance is low and decrease when its abundance is high. Instabilities — the explosions, the collapses, and other seemingly anomalous behaviors — are caused by deterioration of the signal's fidelity.

The issue of population stability is important to our discussion, but the term *stability* has various interpretations. A common definition relates to ecological stability (see, for example, Nisbet and Gurney, 1982: 10; also compare May, 1973, and Emlen, 1973: 360–362). An ecologically stable population persists for many generations; an unstable population, only several generations. The ecological-stability property is theoretically important because virtually all populations are, as a practical matter, ecologically stable; the paucity of data and their relatively large variability make it difficult to detect trends, which might in any event be temporary.

The apparent universality of ecological stability suggests that understanding how it is maintained is an important research topic. Because it is universal, consideration of other properties that can be used to discriminate among empirical changes in population abundance is necessary. These properties concern observed variability. The observed variability can be partitioned into consistent changes in the mean abundance of the population (for example, trends or possibly nonrandom oscillations, such as might be evident in a population with Leslie-matrixlike dynamics) and into short-term stochastic variation that would be evidenced by fluctuations about the mean population abundance. Ordinarily, if trends exist, their effects can be removed, leaving the stochastic variation unaffected by temporal changes in average population abundance. Thus the dynamic behavior of populations can be classified in terms of whether or not trends exist and in terms of whether or not there has been a consistent change in the nature of the stochastic variation. If the stochastic variation has not changed—that is, if it can be assumed to be generated by a sampling from a particular probability distribution with specified parameters (for example, from a normal distribution with mean = 90 and variance = 25)—we can say that the variation is stationary (a less strict definition would include specifying only the mean and variance).

The temporal changes in average population values and in the probability distribution of population variability are evident in recruitment time series, making it important to explain how these changes are induced. In discussing the mechanisms that might underlie trends or changes in stationarity, the previous chapters have been developed as somewhat independent essays on recruitment-related topics. This chapter synthesizes its antecedents to develop a model of the population-dynamics process. The model at this point provides a qualitative picture of the general responses of populations to their own densities, to the densities of associated populations or species, and to the physical environment.

Chapter 1 established that, of the intensively studied fisheries of the world, there are no known instances where recruitment variability has been understood to the extent that it can be predicted; the best that may be said is that variability results from a complex of multiple factors. Chapter 2 considered the problem in more detail by examining the recruitment-variation record. Average variation in recruitment has a range of about ten to one, which can be larger, depending on the definition of range. Examination of the recruitment time series revealed that the data are difficult to interpret in the absence of theory. Because the time series at hand was a small sample from a much longer record, we could not determine whether it was typical or atypical. Accordingly, we could not know whether an unusual value in the sample was globally unusual. Thus to understand recruitment variability, we must go beyond techniques that depend only on recruitment data. In other words, theoretical advances are needed. So that thinking about theoretical development would not have to take place *in vacuo*, Chapter 3 sketched the essence of fish-population-dynamics theory. Chapter 4 took account of previous observations, and we determined (a) that available theory addresses only a fraction of events comprising the population-dynamics process; (b) that the theory's utility is accordingly limited to a restricted set of problems conventional to fish-population-stock assessment; (c) that additional important problems dealing with pollution, habitat, and stock-rebuilding strategies require new theoretical development; and (d) that formation of a new theory requires a better understanding of high-fecundity population dynamics and biological oceanography. This improved understanding must derive from an evaluation of the critical linkage between recruitment and parent stock.

In this context, Chapter 5 explored recruitment-stock linkage. Recruitment-stock theory describes the relation of young-fish mortality to young-fish abundance, a simple but often obfuscated idea. The fact that the extant theoretical relationships between recruitment and stock represent a few special cases (that is, the Ricker and Beverton-and-Holt functions) suggests that other special cases might be considered, particularly in addressing the issue of *why* some functions are better than others. A revaluation of the theory is necessary because there seems to be little correspondence between empirical data and theory, a paradox that might be resolved by recognizing that the essential element in the recruitment-stock relationship — the element that maintains population stability — is the curvilinearity in either the α-function or the γ-function. Hence the variability might be thought of as a separate issue. Both the nature of the curvilinearity and the variability can be addressed through study of each component that contributes to recruitment-stock variability.

It is also important to study the linkages among these components. In this regard, the Paulik diagram is useful. The diagram shows how density dependence in some stages can override a lack of density dependence in other stages; how variability may exist in particular stages as induced by conditions that may occur in only a single year or in a certain temporal sequence; and how some life-history stages may be more important than others in generating recruitment-stock variability. The partitioning of the recruitment-stock process into components organizes the examination of transfer functions or transformations from one stage to another — mature-female biomass into eggs, eggs into larvae, and larvae into recruits. In fact, Paulik-diagram partitioning suggests separate in-depth consideration of adult-related and larval-related transfer functions.

Accordingly, Chapter 6 addressed adult-related processes in the transformation of female biomass to eggs, while Chapter 7 addressed the larval-related processes that account for the fate of eggs and larvae. Chapter 6 reflected that, contrary to implications in much of the literature, egg production is an important part of the populations-dynamics process; its dynamic contribution has been masked because many fish-growth studies "average out" reproductive-tissue growth, limiting the utility of the growth analyses for tropho-dynamics studies (but not necessarily for traditional fishery stock assessment). Our analysis suggested that taking reproductive-tissue growth into account shows that mature female fish *do* grow as a function of their density, a usually denied assertion. Placing additional emphasis on egg production, Chapter 6 also indicated that the usual concept of fecundity is too aggregative to be useful for the study of the population-dynamics process. Fecundity involves many more factors than simply the total number of eggs produced each year. In particular, it is important to evaluate how eggs of certain species vary in terms of their capacity to survive, as a function of, among other things, their size, energy stores, and physiology, as well as how the spawning process allocates eggs to sample the time-space of heterogeneous environments.

The transition from the egg-production stage to the eggs-in-the-sea stage represents a remarkable change in setting. Egg production seems more related to events "controlled" by adult fish — the number of prey per adult and the transformation of ingested energy into eggs — whereas egg and larval survival seems more related to events "controlled" by the oceanic environment — the availability of food, predation, and physical factors. In Chapter 7 the relation of egg and larval mortality to the oceanic environment was discussed. We reviewed the history of research on larval mortality and began to develop a conceptual model of the complex causes of egg and larval mortality. We stressed the inseparability of larval nutrition, predation on larvae, and the larval environment. The need for new conceptual models can readily be appreciated, as only a few years ago egg and larval mortality was basically treated as a problem of only one aspect of larval nutrition (that is, the quantity of food required for larvae to survive starvation death without respect to the nutritional quality of food).

In beginning to think about alternative models, we observed that most events associated with egg and larval mortality had a feature in common: they involved prey-predator interactions. We noted that these interactions occurred not only between predators and eggs and larvae but also between larval fish and their forage resources. Moreover, the quality and quantity of these trophic transactions were clearly affected by the physical environment.

The larval predator-prey milieu is too complex to understand in its totality at present. The best we can do is to examine various metrics that provide a common thread linking prey-predator interactions in the plankton. The generalized distance among organisms in the sea as evaluated by occupancy volumes is useful because the volumes can take into account both trophic interactions and the effects of the physical environment on these interactions. This approach enables us to study causally proximal relations of fish larvae to the physical environment rather than examining causally distal events subject to considerable intervening variation. Tying these notions together suggests the substance and form of a population-dynamics process.

Components of the Process

The recruitment-stock relationship has generally been considered to be equivalent to the population-dynamics process for marine fish. We can now see that this relationship is but a special case of the population-dynamics process, as the recruitment-stock notion implicitly suppresses the relation of recruitment and stock size to the environment, and the role of egg production and events intermediate between egg production and recruitment. This is not to say that the recruitment-stock relationship is wrong or has little utility; rather, it needs to be placed in a broader setting that accounts for and synthesizes concerns for both stability and fluctuation in fish-stock abundance in terms of density-dependent and environmental effects, moving toward explaining the paradoxically large variance and the nature of curvilinearity in recruitment-stock relationships.

The broader setting has several components that, taken together, constitute the population-dynamics process. The process includes a stability-inducing component associated with the species of interest and its trophic status, and a stability-deteriorating component that is more a function of the environment, external to but affecting the trophic interactions of the species of interest. The stabilizing component consists of population-regulation modules, whereas the destabilizing component results from distorted signals on population abundance.

Density-Dependent Regulation Modules

Density-dependent regulation modules are key components in the population-dynamics process. Each module causes the population to increase or decrease and consists of intrinsic and extrinsic mechanisms. Intrinsic mechanisms are primarily related to fish physiology or behavior in the population of interest; extrinsic mechanisms are primarily related to the environment. Egg production is an example of an intrinsic mechanism, controlled by the physiology of the fish. Clearly, egg production is affected by a "food signal," which is in a sense extrinsic; nevertheless, proximal control resides in the fish itself. Predation on larvae is an example of an extrinsic mechanism, as it is arguably more a function of the density and behavior of predators than a function of the prey. Again, the definition is a matter of degree. For example, the intensity of predation may relate to the growth of larvae or to the nature of larval evasive behavior.

The distinction between intrinsic and extrinsic mechanisms is important because their operational modes are different. Intrinsic mechanisms such as egg production, when intensified, tend to enhance population increase. In contrast, extrinsic mechanisms such as predation on larvae, when intensified, tend to dampen population increase. This distinction reflects a certain symmetry between enhancing and dampening mechanisms, characterized by a never-ending, stability-inducing "struggle" between them.

Obviously, the stabilizing operation of enhancement-dampening mechanisms requires an increase in enhancing activities when the population is at a low level of abundance, and vice versa. Similarly, enhancement-dampening mechanisms require an increase in dampening activities when the population is at a high level of abundance, and vice versa.

How does the system "know" that it is at a high or low level? Two signals are used to turn on enhancement or dampening mechanisms. The first is the food signal. When the population is at a low level of abundance, there is more food per individual, a circumstance that turns on the density-enhancing mechanisms in terms of increased growth and fecundity. The second is the predation signal. When the population is at a high level of abundance, it enhances predatory control either instantaneously or on a longer time scale, even possibly inducing several generations of predators to increase in abundance.

The idea that signals control the operation of the regulatory modules opens the possibility of analyzing signal transmission and associated noise in ecological systems. Although the Shannon index (Shannon and Weaver, 1959) is used in ecology, it does not appear to be used in its original context. Further, the food and predation signals envisioned here certainly involve a transfer of energy, but they also involve a transfer of information. Thus some parts of the ecosystem may be more sensitive to energy dynamics, and other parts to information dynamics.

Modules Corresponding to Life-History Stages

Marine-fish populations have four life-history stages: egg production, egg, larval, and juvenile. Correspondingly, there is an egg-production module, an egg module, a larval module, and a juvenile module, which serve as density-dependent population controllers, turning on and off population-enhancing or population-dampening mechanisms, depending on food or predation signals.

Egg-production module. The operation of the egg-production module, where primary control appears to be intrinsic density enhancement, was discussed in detail in Chapter 6. It was pointed out that as population decreased, the food density of juveniles and adults increased, accompanied by an increase in size at maturity or a delay in age of maturity, as well as by a reduced per-unit-weight fecundity. Population collapse may be related to events associated with the egg-production module. Some rather circumstantial evidence suggests that collapsed populations may exhibit a decline in egg production per unit biomass, such as might occur if the system as a whole were highly productive and the population of interest at a high level of abundance. The system then declines in productivity, but the population of interest is still at a high level of abundance — it rapidly overgrazes its resources and blocks the transmission of density-enhancing signals to the egg-production module.

Consider, for example, the circumstantial evidence associated with the decline of the Peruvian anchoveta. For the anchoveta, the "Report of the Consultative Group on the State of the Stocks of Anchoveta and Other Pelagic Species" (Lima, July 1977) noted that through the late 1960s the spawning-stock size was maintained at a level of about 15 million tons and a high recruitment was in evidence. The spawning in September 1971, however, resulted in weak recruitment, noticed in early 1972. Indeed, the recruitment from this year class appeared to be one-eighth the recruitment in the late 1960s.

The El Niño of early 1972 caused — as is typical — warmer waters, which drove fish to concentrate in cooler waters close to shore. This change made the nominal unit of fishing effort more effective. The poor recruitment and intense fishing reduced the spawning stock in September 1972 to about 2 million tons.

After 1972, catches were somewhat less than the moderate recruitment, and the spawning stock averaged about 4 million tons until 1975. In 1975 the recruitment conditions apparently improved, and the spawning stock increased in early 1976 to about 10 to 12 million tons.

This optimistic note was dashed, however, when the El Niño and its warmer waters reappeared off the Peruvian coast in 1976. Again, the fish concentrated close to shore, but feeding conditions were evidently unfavorable since the normal diatom/zooplankton mix was replaced by dinoflagellates in unusual abundance. The fish's growth was reduced; they were lean and had a low fat content. The September 1976 spawning from these fish was "disastrously low, not more than half a million tons, or well below one-tenth of that of the late 1960's . . . additional evidence of the unusual characteristics of the anchoveta stock was that in September 1976 the length of first maturity was 8–9 cm, a few centimeters lower than normal."

The decline of the anchoveta in association with the two El Niños is interesting because the responses of the fish to conditions were apparently different. In 1972 the fish were especially large, but the proportion of mature females during spawning time was particularly low. The fact that these fish may not have spawned is evidenced by the failure of their fat content to decline (see *Report on Peruvian Anchovetta*, 1974), as is typical for clupeoid fish at spawning time (see, for example, Perkins and Dahlberg, 1971). In addition, the number of eggs was greatly reduced. The conditions in 1977 appeared to be somewhat different, in that stock size was at a low level, growth was poor, and dinoflagellates dominated the plankton. The low fat content suggested that the fish were in spawning condition, but the apparently low nutritional level may have affected spawning capability. There is no evidence to indicate at this point whether the poor recruitment was related to spawning or to events during the egg and larval stage.

The per-individual production of reproductive biomass also seems to have declined for the California sardine (see Figure 6.3). It is interesting to speculate how this event might be related to a perturbation in the adult-sardine food signal. Perturbations in the adult food signal, particularly when the adult population size is very large, may be a key element in the onset of stock collapse in the absence of increases in fishing mortality, as discussed in Chapter 4. These examples, of course, beg the question as to why collapsed populations do not again increase. There is no answer to this question, but it is tempting to speculate on a number of possibilities, including a modified genetic structure of the stock (see, for example, Myers and Krebs, 1974).

Egg module. Moving from the egg-production module to the egg module, control seems to shift to extrinsic density dampening. Of the extrinsic mechanisms, egg predation seems to be an important cause of egg death, even though it is not clear to what extent egg predation is density dependent. Density-dependent egg predation requires that predation efficiency change with the absolute abundance, local density, or variance of egg density. As implied earlier, mechanisms of density-dependent predation on eggs relate to characteristics of either the eggs or the predators.

The simplest examples of predator-prey interaction relate to the numbers of egg predators and eggs. Standard observations on predation are applicable. In some instances the predators may eat any number of eggs greater than some fixed number. This finding implies that the fixed number of eggs not susceptible to predation is somehow sheltered from predation, whereas any greater number is exposed to predation. Such a situation might be most easily visualized in the case of demersal eggs that are superimposed on one another, the bottom layer being most protected from predation (but not necessarily from other causes of death). Another alternative is that the predators take no more than a fixed number of prey, implying saturation or satiation effects. However, it would be unlikely for predators to become saturated if in fact food in the sea is as scarce as is sometimes reputed. Also it is possible that the saturation threshold increases as food abundance increases. These possibilities are counterindications of the saturation phenomenon, a particular kind of density dependence.

Beyond these relatively simple concepts, more subtle possibilities are associated with density-dependent predation on fish-egg distribution. Consider, for example, a large and small stock of the same species. As shown in Chapter 6, both could produce roughly the same total number of eggs. The eggs from the larger stock would be partitioned into smaller, more numerous "packages" than the eggs from the smaller stock. The different-sized packages of eggs suggest that eggs from the larger stock would have a more (or less?) patchy distribution than the eggs from the larger stock. If efficiency of predation on eggs is governed by patchiness of egg distribution, then adult-population density generates egg patchiness, which is a density-dependent effect.

Density of fish eggs, and variance in the density, may itself be a function of predation. Just after spawning, fish eggs are presumably at their greatest density. Density is reduced by diffusion and mortality; Pielou (1977) discusses how predation might alter the probability distribution of prey. Now, if mortality is caused to some extent by predation, and if predation is density dependent, then the absolute abundance, density, or variance in egg density is density dependent and in turn affects density-dependent predation.

Some indirect evidence points to density-dependent egg mortality from predation. As Ricker (1954) suggested, cannibalism is the most natural candidate for density-dependent early mortality of young fish (or eggs). When density-dependent egg predation occurs, the relation of recruitment to stock should be "dome shaped," and dome-shaped recruitment-stock curves do exist. Several different mechanisms, however, could induce the dome shape, as suggested in Equation (5.15).

But this possibility is part of the problem. Can we examine a recruitment-stock curve and conclude that it actually represents the relation of recruitment to stock for a relatively long time and, further, can we make inferences on the events intermediate between spawning and recruitment? For example, space for egg development is limited, in that there must be limiting densities of eggs or specific limited locations where egg survival is possible, a concept particularly abstract in regard to pelagic eggs living in a three-dimensional setting, but not for demersal eggs living in a two-dimensional plane. Presumably at relatively low demersal egg numbers, the numbers of larvae produced will be proportional to the number of eggs produced, a situation consonant with either the Beverton-and-Holt or the Ricker form of the curve. As egg densities increase, however, two forms of egg crowding develop. In the first, the areal extent of suitable substrate is occupied by eggs; and in the second, the eggs become locally crowded when superimposed on one another. Under conditions of relatively *high* egg production, conceivably there will be a high initial mortality proportional to the number of eggs, producing a Rickerlike curve. But if egg production increases even more, so that it is relatively *very high,* then egg mortality will not be proportional to the number of eggs produced, thus generating a Beverton-and-Holt form of curve. This example again points to the difficulty of generating curves based on empirical data, as the shape of the curve can depend on the distribution of the independent variable and not necessarily on any particular relation of recruitment to stock (see Rothschild and Mullen, 1985).

Hence it does not appear that the egg module possesses a particular density-enhancing mechanism, but density-dampening mechanisms are evident. There continue to be fine points, however, that require resolution. In some instances the nature of density dependence is not clear. On the one hand, the large numbers of dead eggs that are sometimes found for some species might result from density-independent factors, such as agitation from wind action (Rollefsen, 1932; but see also Pommeranz, 1974). On the other hand, the large numbers of dead eggs might be caused by a density-dependent nutrition syndrome in the spawning females or by inadequate fertilization that might be due to an unbalanced sex ratio at spawning, a possible result of the application of unbalanced fishing mortality to males and females.

Larval module. The larval-regulation module, where control is extrinsic density dampening, seems more complex than either the egg-production or the egg module, because density-dependent regulation at the larval stage involves interaction between two very different density-dependent events: predation on larvae and predation by larvae on their food sources. As indicated in Figure 7.2 and discussed in Chapter 7, it is clear that these two processes cannot be considered separately, as has often been the case. In fact it is the interaction between predation on larvae and the nutritional state or health of larvae that is of critical interest. That is, does predation become more or less effective, all other things being equal, when larval food resources are increased or depleted?

These questions are difficult to answer, as there is but scant knowledge of the actual causes of larval death. Virtually no population-dynamics-related research has been done on predation of larvae, and the research on larval-fish feeding has been often too narrow in scope, or muddled. Of particular concern is the fact that larvae and postlarvae cannot be treated as if they were ontogenetically identical, as has been the practice or implied by considering only larvae of a narrow size or age range in some experimental work. Considerable ontogenetic changes in larval morphology, physiology, and ecological setting are well known. Major concerns are *ontogenetic changes* in (a) larval density, (b) ability to feed, (c) ability to escape predation, and (d) food preferences or habits. Paralleling these concerns is the need for a broader understanding of the nutrition of larvae and their predators, as well as the need to place in perspective the implications of conducting laboratory experiments under conditions where survival rates far exceed those known to occur in the sea.

The extent to which starvation and predation mortality are collectively density dependent is somewhat problematical. Three sorts of observations cited in Chapter 7 may be relevant: (1) estimates of larval mortality rates in the sea, (2) estimates of larval mortality implied by the ratio of healthy to malnourished larvae in the sea, and (3) the number of recruits relative to the number of larvae. In addition to providing some suggestions on the possibilities of density-dependent larval mortality, these observations also warrant additional comment on the operation of density dependence and the relation of larval to juvenile mortality.

One might think that the mechanisms of density dependence operate according to a neat deterministic function where increments or decrements in density always result in corresponding increments or decrements in density-dependent effects. Arguably such a model may exist and be masked by measurement error. But the data suggest (see Figures 6.6 and 7.9) that there are no density-dependent effects over much of the range of the data, the density-dependent effects being observable only at unusually high densities, which by definition are rare. In other words, an alternative model to simple density dependence is one showing that density dependence operates only at extreme densities. Since extreme densities are rare, the accrual of sufficient observations of density-dependent effects from typical time series will be difficult.

It appears, then, that the larval module might be a source of positive feedback, particularly at low and intermediate population levels. Even at these levels the population distributes extremely large numbers of eggs according to the sampling processes implied by Equation (6.2). When the population is at an extremely high level, however, all the "niches" have been sampled, and predation on either nutritionally competent or malnourished larvae takes its toll. It does appear, though, that control at the larval stage is largely due to extrinsic density-dampening mechanisms, which are influenced significantly by the oceanic environment.

Juvenile module. In the juvenile module, control is complex; it seems to shift again from an extrinsic dampening regulation to an intrinsic enhancing regulation. The juvenile stage is important because it is when components of mature-fish biomass are elaborated and size or age at maturity is determined. Although we have a rather clear picture of density-dependent growth in juvenile fish, the effects of density-dependent mortality are not so clear, being supported with relatively little evidence. Lockwood (1980) shows higher mortalities for juvenile plaice that are at higher densities, but as his samples are from different locations, it is difficult to determine whether the effects are simply locational differences or reflective of actual density-dependent phenomena. Perhaps a better example is provided by the gadoid fish, for which various estimates of cannibalistic mortality have been made.

Some researchers have raised the issue of the relative effects of young- and juvenile-fish mortality in terms of population regulation (for example, Sissenwine, 1984). In particular they have noted, implicitly, that the early life history may be partitioned into two stanzas, one of high mortality and brief duration and another of low mortality and relatively long duration. That is, the number of fish that die is a function not only of the instantaneous mortality coefficient but also of the length of time over which it operates. Since the length of the second stanza is relatively long, the product of second-stanza mortality rate and time is inferred to be large and hence to have an important effect on regulation. In actuality, this model is the same as the Beverton-and-Holt interpretation of the Ricker model (Chapter 5). Moreover, the recruitment-stock calculus, as pointed out in Chapter 5, generally gives the interrelationship between numbers of younger fish and older but prerecruit fish. Nevertheless, we can obtain some idea of the relative magnitude of $M_1 T_1$ and $M_2 T_2$. To examine the assertion, let $M_1 T_1$, $M_2 T_2$ represent the mortality-time products for the first and second stanzas, respectively. The implication is that $M_1 \gg M_2$, but that $M_2 T_2 > M_1 T_1$ because $T_2 \gg T_1$. To test the assertion, it would be reasonable to suppose that $M_1 = 0.1$ per day for 30 days; then $M_1 T_1 = 3$ (equivalent to an annual instantaneous rate of about 36).

Now a reasonable range of second-stanza natural mortality might be $0.5 \le M_2 \le 1.0$. Thus for the condition $M_1 T_1 = M_2 T_2$, we would require $3 \le T_2 \le 6$ years, which seems to be about the right magnitude for recruitment time. Let N_0 be the number of fish alive at the beginning of stanza 1; N_1 the number of fish alive at the end of stanza 1; and N_2 the number of fish alive at the end of stanza 2. If $N_0 = 10^{12}$ and $M_1 T_1 = M_2 T_2 = 3$, then $N_1 = 10^{12} \times e^{-3}$, and $N_2 = 10^{12} \times e^{-6}$. Hence the absolute rather than the proportional number of deaths during the first stanza is about twenty times greater than the number of deaths during the second stanza; that is, with these reasonable numbers the scope for variation in mortality appears to be far greater in the first stanza than in the second stanza.

Integration of Modules

Thus each life-history stage can be characterized by a density-dependent regulation module. The four modules are linked together in Figure 8.1. The modules, taken together with the signals they receive or transmit, constitute the population-dynamics process.

Collectively the modules provide system redundancy and ensure that the system will be highly stable; if regulation at one stage fails, the next is likely to provide a more intense compensatory response. This fail-safe redundancy is further enhanced by the degree of independence among the signals received or transmitted at each stage. At each life-history stage the fish eat different food items, so four independent signals are transmitted to the population on its own abundance. Because the signals are independent, ambiguous or equivocal transmission at any one stage is likely to be compensated for by accurate signals at the other stages. Similarly, the population-abundance signals transmitted to predators are also independent; different kinds of predators feed on fish at different life-history stages; hence these signals also have a degree of fail-safe redundancy. For example, (a) egg production is dependent only on the food of juvenile and adult fish; (b) egg mortality is dependent largely on predation on eggs, which could be related to density of adult fish or to other predators on eggs, which are unlikely to be related to the food resources of juveniles or adults; and (c) larval mortality is dependent on the complex interaction of predators on larvae and their prey. Although there may be common elements between the predators of larvae and the predators of eggs, the connection between predators of eggs and larval food seems weak, if existent at all, and predation mortality of juvenile fish would certainly seem to be induced by larger predators than those that feed on eggs and larvae.

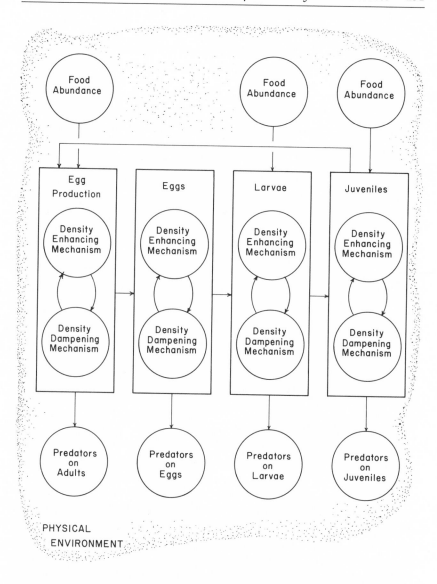

Figure 8.1. The population-dynamics process. This particular hypothetical population consists of four stabilizing modules, each consisting of density-enhancing and density-dampening mechanisms. Food abundance transmits information to the fish population on its own abundance affecting the density-enhancing response. Population density is transmitted to predators. If predators change their efficiency as a result of the signal, then predation is density dependent. Note that redundancy is suggested, inasmuch as food resources and predators are different at each life-history stage. Note that in the multiple-population context predation and food signals are identical.

Note that the time phasing of food and predation signals may be quite different, thereby introducing hysteresis into the system. The food signal is virtually instantaneous, whereas the predation signal may operate on an instantaneous time frame but may also result in turning on the food signal of fish predators for several generations, possibly creating systematic oscillations. The combination of fast-track and slow-track signals is, of course, characteristic of systems that exhibit catastrophes.

Viewing the integrated process enables us to consider a particularly critical problem, namely, the relation among species in the ecosystem, or the so-called multispecies problem. The process depicted in Figure 8.1 represents the control system for a single species, but it obviously contains linkages to other associated species in terms of the trophic interactions implied by the food and predation signals. Because of the high degree of compensation in the system, the decline or increase of one species does not necessarily generate a decline or increase in another *associated* species, *even though* a large amount of caloric energy is involved in the trophic transaction. Further, it is easy to see that among the many opportunities for trophic transactions (eight possibly independent food or predation signals), small energy flows could have large effects. Thus the interaction of prey and predator as posed in most mathematical models is not particularly helpful in understanding these systematic interactions. Other models are needed.

Variations and Fluctuations

If the population-dynamics process is such a highly redundant self-stabilizing system, why then, apart from random measurement error, do we find fluctuations, instabilities, population collapses, and population explosions? The answer, independent of changes in the genetic constituency of the populations, lies in the signals. As long as the signals are normally transmitted, the population remains stationary. But if the fidelity of the signals deteriorates because of increased noise or other factors, then population stationarity can be modified.

Without at this point identifying specific causes of noise, the existence of signals and deterioration in their detectability as a result of noise suggest some interesting properties of the population-dynamics-process control system. The first is that if overall stability is conferred by a control system characterized by several stages of independent control, then control of some populations is inherently more "stable" than control of other populations. The efficiency of control depends on the probability that the controlling system recognizes unusual population abundance. This probability is a function of, among other things, (a) the probability that the unusually high or low abundance is detected at *each* stage and (b) the number of controlling stages or modules. Thus an unusually large year class could simply be the result of the system's failure to "recognize" at each stage that the population was at a high level of abundance, an unlikely but possible event. A second interesting aspect relates to observations suggesting changes in ecosystem character—for example, the Russell cycle (Cushing, 1982), the gadoid outburst, and flip-flops in species composition, such as those reviewed by Daan (1980). In these instances the signals from the effectors must have been sufficiently broad to influence nearly the entire spectrum of receptors; surely these broad signals must relate to environmental changes that tend to affect all organisms, such as changes induced by factors associated with primary or near-primary production, or to physical features of the environment. Put another way, such fundamental and broad-based changes tend to influence each stage, though with possibly different time lags, until eventually the independence among the signals at each stage erodes, the old "stable" state is overcome, and a new "stable" state is generated.

Finally, we have tended to assume that receptor specifications are constant, that is, that predator and prey behavior and genetic constituency in the population of concern are unchanging. This is clearly a great simplification, as it is likely that populations may change in very fundamental ways so that even though signals and their associated noise remain the same, the nature of the receptors changes. An example might be genetic changes in a stock that are evident in physiology but perhaps unnoticed because of lack of accompanying change in morphology, or temperature changes that affect the metabolism of predator or prey, or both, differentially so that even though the signals do not change, the reactions to the signals change.

Given this general background on factors influencing signals, it is possible to discuss some specific mechanisms. These mechanisms may affect the stocks at a single stage or at several stages in concert. Consider first the food signal. The food signal operates such that when the population is abundant, there is less food per individual and reproductive processes are not enhanced, and vice versa. It is easy to see how the food signal may become distorted. Suppose the juvenile or adult population of interest stays constant, and food abundance varies. When food is abundant, the population "thinks" its abundance is relatively low and increases its reproductive efforts; when the abundance of food declines, the population "thinks" it is at a high level of abundance and decreases its reproductive efforts, possibly generating a stock collapse. But to what extent can food abundance actually vary? The answer obviously depends on the food habits of the predator and, in particular, the number of species that it eats. Variation in predator-perceived forage abundance must be dampened as the number of species of forage organisms increases. Moreover, as the total abundance of forage decreases, predators are likely to switch to other species, further dampening the variability in total forage ingested, although quite possibly increasing the energetic or other nutritional costs associated with a "suboptimal diet."

The forage mix appears to be more stable than the abundance of the population of interest. The buffering of the forage mix results from (a) its multispecies nature (assuming that the variability of the species taken together is less than the variability of individual species) and (b) the fact that the species in the forage must have compensatory mechanisms, just as the population of interest has compensatory mechanisms (this is reminiscent of May's 1973 discussion of complexity versus stability, in which he argues that complexity does not necessarily ensure stability). While a bit circular, it could be argued that observations of density-dependent somatic growth in juvenile fish and density-dependent reproductive growth in adult fish suggest that the food signal may be relatively more stable than dynamic changes in the population of interest. If, in fact, the multispecies nature of food signals stabilizes the signals, then the major cause of signal destabilization appears to be the physical environment.

Additional, admittedly circumstantial, evidence regarding variations in food-signal fidelity is derived from the fact that while density-dependent growth may obtain on the average, it does not exist in every year (cf. Figure 6.6). We can infer that (a) the forage is on average stable, but years when the density-dependent pattern is not followed are years of exceptionally high or low forage; (b) the forage is constant, but physical factors affect its availability; or (c) a combination of (a) and (b).

Physical factors can contribute as much as biological factors to food-signal variability. To put this in perspective, consider that the food signal is influenced by (a) the physiology of the predators and the prey — that is, their requirements for nutrition and their reactions to forage, given their state of hunger — and (b) the generalized distance between predator and prey. Recall from Chapter 7 that generalized distance D_i-metrics may be used to parameterize predator-prey interactions in the sea. The establishment of the generalized distance metric permits consideration of the specific effects of the cardinal physical variables — motion, temperature, and irradiance — on the food signal.

Velocities and accelerations in the environment are important physical features affecting the generalized distance between predators and prey in the sea, as motion affects the position and acceleration of individual plankters as well as the energy expended by larger organisms to maintain some particular position or set of positions against the flow of water. As pointed out in Chapter 7, motion affects not only the prey-predator distance but also the exposure of organisms to the other primary physical variables, temperature and irradiance. Interpretation of the effects of motion is extremely difficult, because motion operates on a range of scales from Rossby waves to small-scale turbulence (see, for example, Woods, 1977). To simplify the discussion, two scales of motion can be considered: an advective scale that translocates a mass of plankton, keeping the position of each plankter constant relative to that of every other plankter, and a microscale or fine-scale motion that affects the relative positions of plankters to one another. Advective-scale motion primarily affects the exposure of organisms to other physical features, and microscale motion primarily affects short-term prey-predator inter-actions.

Advective changes (that is, drift) have been considered important since Hjort's time. It is clear that eggs and larvae require salubrious conditions for nutrition and low predatory stress to survive. These conditions are a function of, among other things, geographic position. Since eggs and larvae are advected along with the masses of water in which they are entrained, interannual differences in the velocity or direction of motion could provide for unusually good or poor nutritional conditions for lar-vae as well as for predators on eggs or larvae. Microscale and fine-scale velocities, however, have a more subtle effect; conceivably they might even affect the success of fertilization and predation on eggs and larvae, as discussed in detail in Chapter 7.

Because the effects of motion are intertwined with the effects of tem-perature and irradiance, we must separate effects that relate primarily to motion from those that relate primarily to temperature and irradiance. Higher temperature, as is well known, generally speeds up the physio-logical processes of fish, but in some instances the temperature can be too warm, decelerating physiological rates. As "optimal" temperatures for various physiological processes vary, increases or decreases in tempera-ture place fish or invertebrates that are in optimal temperature in subopti-mal conditions, and vice versa. It might be argued that changes in ocean-temperature patterns at particular locations are quite small and therefore unimportant. In fact, however, reactions to temperature changes that might be barely, if at all, observable in the laboratory could be quite important in the sea; the effects of these slight temperature changes in the field operate on *very large* samples (for example, 10^{12} eggs) and hence could have a most significant effect on the population as a whole.

The effects of irradiance variability have barely been explored. As pointed out in Chapter 4, the existence of the average fish egg or larva is short indeed, and a series of cloudy days could affect the feeding capabilities of visual-feeding fish larvae as well as larval predators. An interesting asymmetry occurs in the effect of irradiance on feeding: among the trophodynamic relations of fish larvae, their prey, and predators on fish larvae, some of the involved organisms are visual feeders, or attracted to light, whereas other are not, suggesting that unfavorable light conditions for visual feeders would enhance the position of nonvisual feeders in the trophodynamic system.

Finally, irradiance is known to affect the time of spawning; since the time of spawning is known to vary considerably within the earliest and latest dates of spring, it is of some interest to determine how this variability is affected by irradiance. An attractive model for the spawning of batch or serial spawners in upwelling areas is that the food signal controls the magnitude of spawning just as it does in nonserial spawning. But in serial spawners the food signal controls timing and the magnitude of spawning pulses. The food signal is generated by a combination of upwelled nutrients and appropriate irradiance conditions; spawning continues until the pulse of food is utilized, and then it becomes latent until the next food signal.

Hence operation of extrinsic and intrinsic mechanisms depends on signals from the environment. The signals have various properties, the most fundamental being their strength and their associated noise. The strength of the food and predation signals is rather simply expressed, even though nuances and eventual complications must be taken into account. In addition, numerous features in the environment invest these signals with noise, thereby creating responses from the population that appear to be dynamically unusual.

Implications of the Process

The population-dynamics process — the process that maintains ecological stability — is intended to explain variations in the numerical abundance of populations. The population-dynamics-process model is in a sense a static model because it addresses the behavior of the population as a fixed genetic entity and does not take into account the important issue of how the species or population-specific stabilizing mechanisms change with time as a function of natural selection. The model is an alternative to the recruitment-stock model; it reflects more explicitly the environmentally responsive, stability-inducing control system; and it places the gaps between existing fish-population-dynamics theory and the population-dynamics process in clearer relief.

The population-dynamics process may be viewed in the context of the many species in the ecosystem or in the context of a single species or population of particular interest. In the multispecies context the process shows how each population is linked to every other population through the nexus of trophic transactions, which in some cases are proximal and in others distal to the population of interest. In the single-population context, the process shows how each population's abundance variations are insulated from those of other populations. The virtual independence of abundance variability among populations is due (a) to the properties of the regulatory module in which the effects of density-enhancing mechanisms balance the effects of density-depressing mechanisms; (b) to the redundancy of the regulatory modules that serve to correct regulation error generated in precedent modules; and (c) to the fact that food and predation signals are inherently stable in the sense that predators and prey generally comprise several populations, and any large variation in one of them would tend to be averaged among the variations of the entire suite of predators or prey, minimizing the effect of an extreme variation in any single prey or predator population.

Describing the ecosystem as a set of population-dynamics processes alters significantly the way that we think about ecosystems. The trophic transactions and the bulk transfer of energy from one population to another have diminished significance by themselves. This is because even though population *A* feeds on a set of populations *B*, changes in the abundance of *A* do not necessarily mean that the populations in *B* will change in abundance. A population in *B* will change in abundance, all other things being equal, only when its compensatory mechanisms are exhausted. This observation challenges models in which the abundance of one species is linked directly and instantaneously to the abundance of another species; the observation directs attention toward physical and chemical variability as affecting the highly stabilized food signals.

This is not to say that energy transfer among populations is not an important regulatory mechanism. Rather, accounting for energy flux in terms of the difference in energy content at two ecosystem nodes is not sufficient to describe the interpopulation or interspecific regulatory mechanisms. It is necessary, in addition, to consider the information associated with the energy transfer. To see this, consider a predator fixed in space with its food moving toward it in a queue. Suppose there is a fixed number of identical food particles and consider the differences in the entropy of the particle distribution if the particles are clustered or if they are distributed randomly in the queue. Consider, too, that the entropy of the particles in the queue can be affected by the physical environment. For example, low-entropy clustering can be transformed to high-entropy random distribution by an onset of turbulence. Hence the predator's nutritional image of prey abundance changes with the spatial distribution of the prey; this phenomenon changes the character of the signals to the predator on its own abundance and initiates short-term reactions in the population that may be of longer-term consequence. As another example, consider that the predator modifies the distributional pattern of the prey. One possibility is that predation erodes the clustering pattern and generates a more random distribution of prey, increasing entropy or actually "consuming" information. As suggested in Chapter 7, it is likely that the morphology, behavior, and physiology of various predators are tuned to an optimum entropy in the distribution of their prey. Consequently a loss or gain in entropy results in a change in the quantity of food ingested — a change in the character of the food signal — even though the actual quantity of food does not change.

The population-dynamics process alters, as well, any opinions that variations in population abundance are associated with either density-dependent or density-independent effects. The phenomenon of density dependence is the quintessential mechanism of population regulation, but its existence does not mean that it can be separated from the influence of the physical environment, which can either enhance or deteriorate the myriad opportunities for density-dependent regulation. Likewise, a consideration of the population-dynamics process seems to suggest that density-independent effects cannot be considered except in the context of the density-dependent phenomenon. Thus, while it is plausible to refer to a density-dependent mechanism, it is difficult to conceive of an *effect* that is either density dependent or density independent.

The concept of the population-dynamics process reemphasizes the fitness of the individual organism and the importance of the population or the species as the fundamental units of ecological consideration. This notion is distinct from advocating the use of ataxonomic criteria to analyze ecosystems dynamics. An ataxonomic approach must integrate populations or species, and it is necessary to demonstrate how the suppression of dynamic characteristics typical of particular populations or species gives insights that are advantageous over those that can be generated by taking account of the traditional taxonomic categories. In one sense the justification for ataxonomic aggregation is lessened by the notion that the abundance of each species or each population seems to be statistically independent from that of any other population. In other words, we could hypothesize that if we were to array the time series of population abundances in a correlation matrix, we would see very few correlations, although there might be temporary associations in trophically proximal populations. If we were interested only in a score of populations, given this situation, aggregation would not be necessary, in that there would be no loss in information if the populations were simply treated additively. Yet, taking account of the individual population-dynamics processes might generate new criteria for aggregation, criteria that involve more than just biomass or energy, an obvious eventual requirement, because of the infeasibility of accounting for every population or species. In fact, admitting the existence of the population-dynamics process reaffirms and perhaps redefines the niche concept and challenges further the possibility that coevolution is a general property of organisms.

The model we have been discussing is essentially different from the usually proposed box models or size-spectrum models (see Ulanowicz and Platt, 1985). The points of difference relate to the methodology for ecosystem partitioning, the use of an information-theoretic setting, and the scope of the model, that is, whether it relates to fundamental phenomena or to monitoring indices that reflect the operational performance of the fundamental phenomena.

With regard to ecosystem-partitioning methodology, most models are partially or wholly ataxonomic. Box models are partially ataxonomic in that scores of species are usually assigned to a single box, and the species may or may not be dynamically similar. In fact, most box models are structured such that the organisms in any one box are of a similar size. The size-spectrum models are completely ataxonomic, taking account of the size distribution of particles in the sea and the flux of particle size. The population-dynamics-process model *is* taxonomic in that it consists of all the population-dynamics-process models in the community appropriately linked by food and predation signals. This is not to say that the population-dynamics processes could not be partitioned, or that multipopulation sectors could not be isolated and studied, or that a particle-dynamic structure could not be derived from population-dynamics-process considerations; rather, any procedure that integrates components of the ecosystem must first take into account the genetically homogeneous units, the populations or the species, and their individual dynamics in developing aggregation or integration criteria.

When we take into account information theory, we see that most ecological modeling efforts use the rate of change of entropy or information to measure system complexity or to demonstrate the gain in entropy according to thermodynamic principles. By contrast, our model suggests the use of information theory, not in the thermodynamic sense, but in the signal-processing sense, providing a methodology to link the physical environment and the population-dynamics processes. The metabolism of ecosystems may be determined as much by information flow (in the signal-processing sense) as by energy flow, challenging the adequacy of all ecosystem models that rely totally on the processing of energy or some other property of biomass.

Consider now whether existing models take into account fundamental biological phenomena or whether they monitor consequences of these phenomena. We can only note that neither box models nor size-spectrum models explicitly take into account the replication or reproduction of their elements. The population-dynamics-process model does. If we view ecology as the study of the interrelationship of organism dynamics and the physical environment, then we have to consider the linked population-dynamics-process models of each population as the ecological engine. This concept does not minimize the importance of other concepts such as the flow of energy, the way energy cycles through the system, or the size distribution and metabolism of particles, which are quite useful as gauges to monitor the performance of the ecological engine. Rather, it says that explicit attention needs to be addressed to the reproductive processes, particularly those that are affected by density dependence. Thus it is important to distinguish between monitoring the performance of a system and understanding and predicting the way it works.

If we believe in the essential elements of the population-dynamics-process model, then we need to reexamine our approach to a number of problems that are common in fish-population dynamics and perhaps in ecology as well. One of these problems is the multispecies problem mentioned earlier. The problem has been generated by what might be called the multispecies viewpoint: if species A eats species B, then species A affects the temporal abundance of species B. Certainly under these circumstances species A will affect the abundance and mortality rate of species B in the short run. But it is not at all clear that species A will affect variations in species B in the long run, as variability in A will first affect the compensatory mechanisms in B before it affects the abundance of B. In other words, variations in A will not influence variations in B unless variations in A are sufficiently large to exhaust the compensatory resources of B. Hence the multispecies problem is not so much a problem of mathematically generalizing single-species population-dynamics models to take into account multispecies interactions, as it is a problem of identifying the biological-physical conditions where the independence among species breaks down, stimulating a change in the fundamental properties of each population or species. These postulates are of, course, tested by fishing, which serves as a proxy for species *A*. Fishing increases the mortality of the stock, but it does not necessarily reduce the productivity of the stock; as a matter of fact, fishing could result in productivity increases.

The problem of isolating the effects of the physical environment on population abundance is another problem of considerable difficulty. The interaction of the physical environment with population variability cannot be understood by referring only to data on the environment and on recruitment, particularly taking account of the population-dynamics processes in the ecosystem. If the system is viewed as a nonlinear negative-feedback system consisting of positive and negative feedback subsystems (as J. Smagorinsky has characterized the ocean-climate system [personal communication]), then our capability for predicting the interaction of the physical environment and the fish stocks will be qualitatively the same as our capability for predicting the general behavior of any such complex systems. The task is indeed difficult. Perturbation of a component in complex systems may or may not yield an immediate effect, owing to the richness of system compensatory or feedback properties. Further, if an effect is not observed immediately, then it could appear at some variable time after the perturbation. Finally, identical perturbations might not always result in the same effect, as properties residing at other dimensions of the system may be changing.

Taking a simplistic view of the fish-environment system yields predictable results. We should not be surprised, for example, when correlations between recruitment and the environment deteriorate shortly after publication of the results (a classic case may be found by reading in concert the papers of Carruthers, Lawford, and Veley, 1951; Gulland, 1952; and Saville, 1959). Observing a simple environmental correlation that exists for only a short time might suggest not so much that there is no relation as that the setting for the relation is dynamic and not appropriately identified. As a corollary, we should be surprised when simple environmental correlations persist. When they do persist, it behooves us to explain how such simple relations override the compensatory mechanisms that must be inherent in the population!

The existence of a population-dynamics process such as that depicted in Figure 8.1 implies that only a small part of environmentally induced variability can be measured if observations are restricted to simple responses to simple environmental signals. Put another way, the simplistic empirical approach sets rather narrow bounds on the relationships that can be contemplated, let alone understood, and explains why there is so little success in predicting the more complex effects of the environment on oceanic biota. The bounds in fact enable us to separate the kinds of causal variation that will be easy to detect from those that will be difficult to detect. The easily detectable relationships involve instances where single environmental variables take on unusually high or low values or instances where environmental variables affect the metabolism of the entire system. In some cases temperature or irradiance changes, for example, are such that the system as a whole tends to be either accelerated or decelerated by inducing a coupling among the food or predation signals of the several life-history stages that are ordinarily independent, resulting in a loss of system redundancy. The components of variance that are most difficult to observe and yet likely to make the most substantial contribution to variation are those involving synergistic interaction among variables, or those with time-lagged responses. In sum, while the distinction is not clear-cut, the detection of relationships between environmental variables and recruitment is easiest when the variables are at unusually high or low levels or when they affect all modules similarly. Events in which single variables are at unusually high or low magnitude must be rare. Likewise, events that drive all modules to react in the same direction must also be rare. This inference suggests that the most common variation results from other sorts of more complex multivariate events.

The population-dynamics process suggests reviewing the loci where regulatory control might operate. As must be evident from Chapter 7, many workers have believed that regulatory control is most intense at the larval stage just after hatching, and consequently density-dependent effects at other stages are minimal. Further, examination of density-dependent control at the larval stage has tended to be restricted to a narrow aspect of larval nutrition and has not, with a few exceptions, taken account of predation or the physical environment. The broader aspects of the problem are exemplified by Rothschild (1981), Hunter (1981), and Rothschild et al. (1982). Considerations such as these suggest reevaluating the causes of larval death and the contribution of stages other than the larval stage to regulation of the abundance of fish populations.

On the cause of larval death, there is sufficient evidence of malnourished larvae in the sea to suggest that continued work on larval nutrition is important particularly with regard to (1) the broader issues of nutrition, (2) the effects of the physical environment on the food signal, and (3) the effects of nutrition on a larva's capability to avoid predation.

With regard to considering regulation at other stages as being effective in controlling the abundance of the population, there is considerable opportunity for density-dependent control at the adult egg-production stage. Indeed, according to the population-dynamics process, control can be and is exerted at all life-history stages; hence the attempt to identify the locus of control may not be a response to a well-formed question, in that a failure in control in any one stage is compensated by relaxed or intensified control at subsequent stages. If we were forced to focus on particular stages, we would observe that the opportunity for numerical rather than percentage variation is considerably greater in larval and postlarval fish than at any other stage. While it is difficult to extricate one's argument from the circular nature of a life history, it does appear that density-dependent control is extremely important at the egg-production stage as well. At this stage there is also an opportunity for large variations in the absolute number of eggs produced. These eggs function to sample the environment for salubrious living conditions. As a sampling process, a reduction in population fecundity decreases the efficiency of the search and increases the sampling variance associated with an egg finding suitable circumstances for survival. Density-dependent regulation seems relatively limited at the egg and hatching stage. The regulation is most intense just after hatching and in the egg-production stage. It is important to evaluate the importance of the juvenile stage in population regulation. Evident density-dependent events occur at this stage; however, in making this evaluation it is necessary to remember the numerical magnitudes of the eggs and larvae occurring at the egg-production, the egg, and the larval stages. Moreover, the fact that a particular stage suffers considerable predation does not mean that such predation is density dependent: if the efficiency of predation does not depend on the density of the juveniles, then clearly it is not density dependent, and it has little direct regulatory influence.

The population-dynamics process provides an opportunity to identify sources of variation that could only be attributed to unexplained variation in the more simplistic approaches. There is little hope, however, that our understanding of its mechanisms will increase through empirical studies of what seem to be almost randomly selected sets of independent variables. A major obstacle to comprehending the process is the extant empirical approach, which has tended to aggregate large numbers of interacting variables, thereby confounding predictions of their influence on the system. The overaggregation issue can be resolved only through considering the individual system components and developing criteria for their reassembly. The overaggregative characteristic of extant models should not be taken to imply that more variables should be included, as this is the difficulty attendant to interpreting large simulation models. Nor does it seem reasonable to think much additional information can be gained from statistical analysis of larger and larger data sets; as the sets become larger and the techniques more sophisticated, the probability of faulty inferences tends to increase. The concern about overaggregation relates to a concern about system structure.

The population-dynamics process suggests emending interpretation of the traditional models. It suggests (a) reinforcing the importance of the relation of the yield-per-recruit model to the recruitment-stock model, (b) reevaluating the production model, and (c) improving the explanation of variation in the recruitment-stock model.

The yield-per-recruit and recruitment-stock models have generally been treated independently, giving rise, for example, to the supposition that stocks can be either "recruitment overfished" or "stock overfished." The population-dynamics process, however, stresses the relation of the yield-per-recruit model to the recruitment-stock model. To reiterate, yield-per-recruit is a function of biomass, and biomass is a function of growth and mortality. As growth increases, egg production increases, thus modifying the recruitment-stock function.

The relation of yield-per-recruit to recruitment and stock is evident from data on changes in North Sea fish populations. Examine, for example, Table 1.1, which capsulizes changes in recruitment and yield per recruit in the North Sea. We can see that not only did recruitment increase in the North Sea, but yield per recruit increased as well, suggesting that the biomass of cod, haddock, plaice, sole, and whiting also increased in a yield-per-recruit sense. One can infer that there was a greater opportunity for the stocks to produce eggs in later years than in earlier years (suggesting the possibility that adult-related phenomena were involved in the outburst). Thus stock overfishing seems generally intertwined with recruitment overfishing and cannot be separated, a conclusion of considerable significance to fish managers.

The production model, we recall, is a particular integration, subject to constraints, of components of both the yield-per-recruit and recruitment-stock models (see, for example, Schaefer and Beverton, 1963). The most important concern regarding the production model is its form — whether for a particular stock the model tends toward a parabolic or an asymptotic shape, for example. Schaefer's (1954) original model of a parabolic steady-state production *must* be at least a nearly correct formulation because a stock cannot maintain a high level of productivity in the face of ever-increasing fishing mortality. Yet, as is well known, most actual fisheries data reflect that the relation of production to fishing mortality tends toward an asymptotic form. The simplest explanation, then, for the asymptotic form is that as fishing effort increases, production increases; that is, as the stock is driven down by increased fishing effort, compensatory growth and possibly reductions in mortality begin to operate, changing the production parameter values and invalidating the underlying assumptions generally associated with production models. In short, the simple relation of inputs to outputs does not obtain. The relation of inputs to outputs, in the context of production, is of course the essence of the population-dynamics-process model; therefore it would be appropriate to identify how changing growth, mortality, and reproduction under specified changes in physical variables affects the relation of rate of change of biomass to biomass.

With regard to the recruitment-stock model, the principal issues involve the large variance associated with theoretical relationships. Although this total variance in the recruitment-stock relation cannot be reduced, apart from measurement error, it can be better explained. The population-dynamics-process model, as well as the discussion in Chapter 7, showed that many events intervene between the magnitude of the stock and the point in time when young fish become recruits. Variation in each event contributes to the total variation seen in the relation of stock to recruitment that appears in quadrant I of the Paulik diagram. In effect, variation in the recruitment-stock relationship may be found in various quadrants of the Paulik diagram. The Paulik-diagram notion of various compensatory events and the mutual instabilities shows that there are ample opportunities for shifting among recruitment-stock domains, a notion recognized by proponents of multistable states in fish stocks (see, for example, Peterman, Clark, and Holling, 1979, or Beddington, 1984). Although the idea of multistable states suggests deterministic relationships, it is conceptually difficult to think of any particular functional relationship, as the actual relationship is likely to be a mixture of events. These domains are what were called "conditions" in Chapter 4, and their identification is a key problem. The process also suggests that the implicitly, if not explicitly, assumed independence among recruitment-stock points may be misleading and that the interdependence among these points may be more important than actually surmised, a feature consonant with the population-dynamics process. (Because of the processes implied by the model, these interdependencies will not be evident simply from an examination of the time-series data.) Another significant source of variation results from the equation of stock biomass to egg production by many researchers. In recruitment-stock theory, the principal interest is in egg production, and, as shown in Chapter 6, stock biomass may not provide a good estimate of egg production. In addition, because of the large number of eggs produced, and the fact that in some stocks the spawned eggs are evidently moribund, it is difficult precisely to define egg production. Thus it is obviously possible to adjust the recruitment-stock relationship for variation in egg production and reduce variance. The population-dynamics process accordingly lends some insights into the large variation associated with recruitment-stock theory. On the one hand, the theory is useful for stock assessment, and its various derivations and consequences are important to think about, as they provide important explanations of recruitment and stock behavior. On the other hand, it is unlikely that conditions remain stable over the length of the data set and that the population responds in the same way to its own abundance and environment. Hence the variation is not relatively large but remarkably small owing to the population-dynamics process.

In addition to analyses aimed at modifying traditional models, the population-dynamics process suggests requirements for their extension, particularly taking account of the need to provide information to support ecosystem management, stock-rebuilding strategies, habitat management, and aquaculture activities. As implied in Chapter 4, a prerequisite to extending the standard models is a better understanding of recruitment variability. The most critical concern is to detect or predict the initiation of a trend or a departure in the stationarity of the recruitment distribution. Stationary recruitment seems to require that environmental conditions remain within certain "normal" bounds. The critical issue is to define the normal bounds so as to be able to detect excesses and to determine the extent to which the excesses have natural or anthropogenic causes, suggesting various tactical or strategic management activities.

Because the system is complex and multivariate, it is necessary to consider normal bounds not only in terms of the values of the variables with respect to some average value, but also in terms of how they are related to their variance-covariance matrix. But the effects associated with these variables are dependent on the magnitudes of the variables and the way in which they vary and covary, as well as on the state of the system. In other words, the system can exist in a number of states, some of which are much more sensitive to environmental change than others, owing to the compensatory properties of the system. Hence it is not a simple matter to determine whether the system has changed. Obviously the separation of natural and anthropogenic effects is even more complex, because not only do we need to detect a change in the populations and in the environment — that is, the subset of elements in the environment that affect the fish — but we also need to detect those components of the population and of the environment that have changed because of anthropogenic activity.

We are led to conclude again that a penetration of these problems must be preceded by a critical dissection of the suppositions that aggregate and hence confound the most sensitive parts of the dynamic system. To place in perspective the overaggregation syndrome, we might find it helpful to recall that at the time when the atom was thought to be the smallest physical particle, our understanding of the physical forces of the universe was negligible. Only when it was realized that the atom consisted of elementary particles could the theory on matter advance sufficiently to permit an understanding of all the forces except gravity. Similarly, fish-population dynamics may be at a stage similar to the field of physics when the atom was believed to be the most fundamental particle. Disaggregation of structures may well lead to new insights into the behavior of the system.

Theoretical revisions and restructuring will require going beyond the study of any particular species of fish. The ecological structures in which the fish live must be reconsidered and redefined. This goal will require an improved understanding of the productivity of the sea, particularly the processes that link productivity at higher trophic levels to productivity at the primary level.

Obviously any consideration must be a simplification; yet it is becoming clear that the energy-flow approaches have already extracted close to a maximal amount of information from the observations at hand.

An alternative approach might be found in a harmonization of the energy-flow, classic population-dynamics, and signal-flow models. The energy-flow models would provide the capability to transfer energy through the system. The classic population-dynamics models would provide the capability to consider discrete parcels of energy in terms of their birth, growth, and death, and in particular the density-dependent, saturation, and amplification effects discussed in Chapter 4. The signal-flow models would take into account the components of the system that transfer information to turn on or off compensatory processes as well as the noise that accompanies the signals. In essence then, the combined models would consider energy flow, but rather than flow *en masse*, flow would be parceled into discrete "particles," each having a particular propensity to reproduce, grow, and die. In addition, the signals and noise could also be accounted for, suggesting the more complex possibility that both signals and noise could be affected by the innate behaviors of the involved organism. Thus we see the conventional energy-flow models as shifting from a Eulerian view of ocean biodynamics to a Lagrangian view, taking into account information-theoretical control of energy transfer.

The study of particle dynamics requires a coordinate system, which considers the distance between or among particles as an index of potential contact and interaction (the D_i-spaces). Sheldon, Prakash, and Sutcliffe's (1972) idea on the size spectrum of particles in the sea is a good beginning point, because the potential swimming velocities and accelerations of organisms are generally related to their size, and so the size spectrum can be generalized to a swimming-speed velocity or acceleration spectrum. The swimming-speed velocity or acceleration spectrum can be transformed to yield a size-specific spectrum of organism motions, relative to one another. The spectrum of contact rates can then be moved closer to a spectrum of trophic interaction by taking further account of ambient motion, temperature, and irradiance.

This formulation broadens the considerations of the D_i-space discussed in Chapter 7. It intensifies our search for a framework suitable to describe the theme of the obvious intimate relation of predator-prey interactions to the physical properties of the sea. It in fact postures the study of variations in year-class strength to reflect on direct physical and biological linkages rather than rely on the endpoints of rather long causal chains.

The need exists for a unifying structure — a biodynamic theory — to study organisms of the sea and a metric to measure the dynamics, which biological oceanographers have discussed for many years (for example, Cushing, 1959, and Riley, 1963). In developing such a structure, we might consider the particles of the sea as analogous to the stars and galaxies of the cosmos. Like cosmologists, we would want to trace the birth, the metabolism, and the death of the particles peculiar to our disciplines. As with the particles of the cosmos, the particles of the sea are abstractions: their dynamics can only be inferred from sensors of relatively limited capability. Hence understanding their dynamics requires exceptional theoretical development. As a practical matter biological problems of the sea are no more or no less complicated than the physics of the universe; hence they too are capable of theoretical innovation, joint focus on critical problems, and development of sophisticated and costly sensors that will permit theoretical validation and elaboration.

Such developments now seem possible in biological oceanography. Static descriptions of the constellations of particles in the sea are now in hand. The next step is to understand the generalized distance among particles and its implications for their trophic transactions. It is in our intellectual grasp and in the range of technological possibility to place ourselves at a point in space-time in the sea and to behold the order in the oceanic universe, the galaxies of living things. By developing theories regarding their future, we can achieve a better understanding of one small but interesting part of our universe, the dynamics of fish stocks.

References

Alverson, D. L., and M. J. Carney. 1975. A graphic review of the growth and decay of population cohorts. *J. Cons. int. Explor. Mer,* 36:133–143.

Anderson, K. P., and E. Ursin. 1977. A multispecies extension to the Beverton and Holt theory of fishing, with accounts of phosphorus circulation and primary production. *Meddr Danm. Fisk.-og Havunders,* N.S., 7:319–435.

Anderson, L. G. 1977. *The economics of fisheries management.* Baltimore: Johns Hopkins University Press.

Andrewartha, H. G., and L. C. Birch. 1954. *The distribution and abundance of animals.* Chicago: University of Chicago Press.

——— 1984. *The ecological web.* Chicago: University of Chicago Press.

Anthony, V. C. 1971. The density dependence of growth of the Atlantic herring in Maine. *Rapp. P.-v. Réun. Cons. int. Explor. Mer,* 160:197–205.

Arthur, D. K. 1977. Distribution, size, and abundance of microcopepods in the California Current system and their possible influence on survival of marine teleost larvae. *Fish Bull.,* U.S., 75:601–611.

Bagenal, T. B. 1966. The ecological and geographical aspects of the fecundity of the plaice. *J. Mar. Biol. Ass.,* U.K., 46:161–186.

——— 1969a. The relationship between food supply and fecundity in brown trout *Salmo trutta* L. *J. Fish. Biol.,* 1:167–182.

——— 1969b. Relationship between egg size and fry survival in brown trout *Salmo trutta* L. *J. Fish. Biol.,* 1:349–353.

——— 1973. Fish fecundity and its relation with stock and recruitment. *Rapp. P.-v. Réun. Cons. int. Explor. Mer,* 164:186–198.

Bailey, K. M., and J. Yen. 1983. Predation by a carnivorous marine copepod, *Euchaeta elongata* Esterly, on eggs and larvae of the Pacific hake, *Merluccius productus. J. Plank. Res.,* 5:71–82.

Bannister, R. C. A., D. Harding, and S. J. Lockwood. 1974. Larval mortality and subsequent year-class strength in the plaice (*Pleuronectes platessa* L.). In Blaxter, 1974.

Baranoff, Th. I. 1918. On the question of the biological basis of fisheries. *N.-i. Ikhtiologicheskii Inst.*, 1:81–128.

Beddington, J. R. 1984. The response of multispecies systems to perturbations. In R. M. May, ed., *Exploitation of marine communities*. Berlin: Springer-Verlag.

Beers, J. R., and G. L. Stewart. 1967. Micro-zooplankton in the euphotic zone at five locations across the California Current. *J. Fish. Res. Board Can.*, 24:2053–68.

———— 1970. Numerical abundance and estimated biomass of micro-zooplankton. In J. D. H. Strickland, ed., The ecology of the plankton off La Jolla, California, in the period April through September, 1967 (part 4), *Bull. Scripps Instn. Oceanog.*, 17:67–86.

———— 1971. Micro-zooplankters in the plankton communities of the upper waters of the eastern tropical Pacific. *Deep Sea Res.*, 18:861–883.

Berg, H. C. 1983. *Random walks in biology*. Princeton, N.J.: Princeton University Press.

Beverton, R. J. H. 1963. Maturation, growth, and mortality of clupeid and engraulid stocks in relation to fishing. *Rapp. P.-v. Réun. Cons. int. Explor. Mer*, 154:44–67.

Beverton, R. J. H., and S. J. Holt. 1957. On the dynamics of exploited fish populations. *Fish. Invest. Lond.*, ser. 2, 19.

———— 1959. A review of the lifespans and mortality rates of fish in nature, and their relation to growth and other physiological factors. *CIBA Fdn. Colloq. Aging*, 5:142–177.

Beyer, J. E. 1982. Models. In Rothschild and Rooth, 1982, pp. 300–328.

Beyer, J. E., and G. C. Laurence. 1979. Modelling growth and mortality of larval herring *(Clupea harengus)*. ICES/ELH Symp./M:6.

———— 1980. A stochastic model of larval fish growth. *Ecol. Modelling*, 8:109–132.

Blaxter, J. H. S. 1968. Visual thresholds and spectral sensitivity of herring larvae. *J. Exp. Biol.*, 48:39–53.

———— 1969. Development: eggs and larvae. In W. S. Hoar and D. J. Randall, eds., *Fish Physiology*, vol. 3. New York: Academic Press.

———— 1971. Feeding and condition of Clyde herring larvae. *Rapp. P.-v. Réun. Cons. int. Explor. Mer*, 160:128–136.

————, ed. 1974. *The early life history of fish*. Berlin: Springer-Verlag.

Blaxter, J. H. S., and G. Hempel. 1963. The influence of egg size on herring larvae. *J. Cons. int. Explor. Mer*, 28:211–240.

Blaxter, J. H. S., and J. R. Hunter. 1982. The biology of the clupeoid fishes. In J. H. S. Blaxter, F. S. Russell, and M. Yonge, eds., *Advances in marine biology*, vol. 20. London: Academic Press.

Blaxter, J. H. S., and M. Staines. 1971. Food-searching potential in marine fish larvae. In D. J. Crisp, ed., *Fourth European marine biology symposium*. Cambridge: Cambridge University Press.

Bowman, A. 1932. The effect on the stock of the capture of undersized fish. The haddock population of the North Sea plateau. *Rapp. P.-v. Réun. Cons. int. Explor. Mer*, 80:1–18.

Boyce, M. S. 1984. Restitution of *r*- and *K*-selection as model of density-dependent natural selection. *Ann. Rev. Ecol. Syst.*, 15:427–447.

Braudel, F. 1981. *The structures of everyday life.* Civilization and capitalism 15th–18th centuries, vol. 1. New York: Harper and Row.

Brett, J. R. 1979. Environmental factors and growth. In W. S. Hoar, D. J. Randall, and J. R. Brett, eds., *Fish physiology*, vol. 8. New York: Academic Press.

Brett, J. R., and T. D. D. Groves. 1979. Physiological energetics. In W. S. Hoar, D. J. Randall, and J. R. Brett, eds., *Fish physiology*, vol. 8. New York: Academic Press.

Brongersma-Sanders, M. 1957. Mass mortality in the sea. In J. W. Hedgpeth, ed., *Treatise on marine ecology and paleoecology. Geol. Soc. America. Memoir*, 67(1):941–1010.

Burd, A. C. 1978. Long-term changes in North Sea herring stocks. *Rapp. P.-v. Réun. Cons. int. Explor. Mer*, 172:137–153.

Burd, A. C., and D. H. Cushing. 1962. I. Growth and recruitment in the herring of the southern North Sea. II. Recruitment to the North Sea herring stocks. *Fish. Invest. 2*, 23:71.

Burd, A. C., and B. H. Holford, 1971. The decline in the abundance of the Downs herring larvae. *Rapp. P.-v. Réun. Cons. int. Explor. Mer*, 160:99–100.

Burd, A. C., and W. G. Parnell. 1973. The relationship between larval abundance and stock in the North Sea herring. *Rapp. P.-v. Réun. Cons. int. Explor. Mer*, 164:30–36.

Bye, V. J. 1984. The role of environmental factors in the timing of reproductive cycles. In G. W. Potts and R. J. Wootton, eds., *Fish reproduction.* London: Academic Press.

Carruthers, J. N., A. Lawford, and V. F. C. Veley. 1951. Fishery hydrography: Brood strength fluctuations in various North Sea fish with suggested methods of prediction. *Kieler Meeresforsch*, 8:5–15.

Christy, F. T., Jr., and A. Scott. 1965. *The common wealth in ocean fisheries.* Baltimore: Johns Hopkins University Press.

Churchman, C. W. 1968. *The systems approach.* New York: Dell.

Clark, C. W. 1974. Possible effects of schooling on the dynamics of exploited fish populations. *J. Cons. int. Explor. Mer*, 36:7–14.

——— 1976. *Mathematical bioeconomics: the optimal management of renewable resources.* New York: Wiley.

——— 1985. *Bioeconomic modelling and fisheries management.* New York: Wiley.

Clark, F. N., and J. C. Marr. 1956. Population dynamics of the Pacific sardine. *Calif. Coop. Oceanic Fish. Invest. Progr. Rep.*, July 1, 1953–March 31, 1955:11–48.

Cochran, W. G. 1953. *Sampling techniques.* New York: Wiley.

Cohen, E. B., M. D. Grosslein, M. P. Sissenwine, F. Steimle, and W. R. Wright. 1982. Energy budget of Georges Bank. *Can. Spec. Publ. Fish. Aquat. Sci.*, 59:95–107.

Colton, J. B., Jr. 1959. A field observation of mortality of marine fish larvae due to warming. *Limnol. Oceanogr.*, 4:219–222.

Coombs, S. H., R. K. Pipe, and C. E. Mitchell. 1981. The vertical distribution of eggs and larvae of blue whiting *(Micromesistius poutassou)* and mackerel

(*Scomber scombrus*) in the eastern North Atlantic and North Sea. *Rapp. P.-v. Réun. Cons. int. Explor. Mer,* 178:188–195.

Crabtree, S. J., and D. M. Ware. 1975. An analysis of some factors affecting the growth rates of Gulf of St. Lawrence cod from 1949 to 1973. ICNAF Res. Doc. 75/98 No. 3584, Dartmouth, N.S.

Crutchfield, J. A., and A. Zellner. 1961. Economic aspects of the Pacific halibut fishery. *Fishery industrial research,* vol. 1. Washington, D.C.: U.S. Department of the Interior.

Csirke, J. 1980. Recruitment in the Peruvian anchovy and its dependence on the adult population. *Rapp. P.-v. Réun. Cons. int. Explor. Mer,* 177:307–313.

Cushing, D. H. 1959. On the effect of fishing on the herring of the southern North Sea. *J. Cons. int. Explor. Mer,* 24:283–307.

—— 1967. The grouping of herring populations. *J. Mar. Biol. Ass.,* U.K., N.S., 47:193–208.

—— 1969. The regularity of the spawning season of some fish. *J. Cons. int. Explor. Mer,* 33:81–92.

—— 1971. The dependence of recruitment on parent stock in different groups of fishes. *J. Cons. int. Explor. Mer,* 33:340–362.

—— 1975. The natural mortality of the plaice. *J. Cons. int. Explor. Mer,* 36:150–157.

—— 1977. The problems of stock and recruitment. In J. A. Gulland, ed., *Fish population dynamics.* New York: Wiley.

—— 1980. The decline of the herring stocks and the gadoid outburst. *J. Cons. int. Explor. Mer,* 39:70–81.

—— 1981. *Fisheries biology: a study in population dynamics,* 2nd ed. Madison: University of Wisconsin Press.

—— 1982. *Climate and fisheries.* London: Academic Press.

——, ed. 1983. *Key papers on fish populations.* Oxford: IRL Press.

—— 1984. The gadoid outburst in the North Sea. *J. Cons. int. Explor. Mer,* 41:159–166.

Cushing, D. H., and J. G. K. Harris. 1973. Stock and recruitment and the problem of density dependence. *Rapp. P.-v. Réun. Cons. int. Explor. Mer,* 164:142–155.

Cushing, D. H., and J. W. Horwood. 1977. Development of a model of stock and recruitment. In J. H. Steele, ed., *Fisheries mathematics.* London: Academic Press.

Daan, N. 1975. Consumption and production in North Sea cod, *Gadus morhua:* an assessment of the ecological status of the stock. *Neth. J. Sea Res.,* 9:24–55.

—— 1978. Changes in cod stocks and cod fisheries in the North Sea. *Rapp. P.-v. Réun. Cons. int. Explor. Mer,* 172:39–57.

—— 1980. A review of replacement of depleted stocks by other species and the mechanisms underlying such replacement. *Rapp. P.-v. Réun. Cons. int. Explor. Mer,* 177:405–421.

—— 1981. Comparison of estimates of egg production of the southern bight cod stock from plankton surveys and market statistics. *Rapp. P.-v. Réun. Cons. int. Explor. Mer,* 178:242–243.

de Ciechomski, J. D. 1966. Development of the larvae and variations in the size of the eggs of the Argentine anchovy *Engraulis anchoita* Hubbs and Marini. *J. Cons. int. Explor. Mer,* 30:281–290.

den Boer, P. J. 1968. Spreading of risk and stabilization of animal numbers. *Acta Biotheor.,* 18:165–194.

Detwyler, R., and E. D. Houde. 1970. Food selection by laboratory-reared larvae of the scaled sardine, *Harengula pensacolae* (Pisces, Clupeidae) and the bay anchovy, *Anchoa mitchilli* (Pisces, Engraulidae). *Mar. Biol.,* 7:214–222.

de Veen, J. F. 1976. On changes in some biological parameters in the North Sea sole (*Solea solea* L.). *J. Cons. int. Explor. Mer,* 37:60–90.

de Vlaming, V. L. 1971. The effects of food deprivation and salinity changes on reproductive function in the estuarine gobiid fish, *Gillichthys mirabilis. Biol. Bull.,* 141:458–471.

Devold, F. 1963. The life history of the Atlanto-Scandian herring. *Rapp. P.-v. Réun. Cons. int. Explor. Mer,* 154:98–108.

DeVries, T. J., and W. G. Pearcy. 1982. Fish debris in sediments of the upwelling zone off central Peru: a late Quaternary record. *Deep-Sea Res.,* 28:87–109.

Dickson, R. R., J. G. Pope, and M. J. Holden. 1974. Environmental influences on the survival of North Sea cod. In Blaxter, 1974.

Dow, R. L. 1969. Cycle and geographic trends in sea water temperatures and abundance in American lobster catches. *Science,* 164:1060–62.

———— 1978. Effects of sea surface temperature cycles on landings of American, European, and Norway lobsters. *J. Cons. int. Explor. Mer,* 38:271–272.

Ellertsen, B., E. Moksness, P. Solemdal, T. Strømme, S. Tilseth, T. Westgard, and V. Øiestad. 1980. Some biological aspects of cod larvae (*Gadus morhua* L.). *FiskDir. Skr. Ser. HavUnders,* 17:29–47.

Elliott, J. M. 1984. Numerical changes and population regulation in young migratory trout *Salmo trutta* in a Lake District stream, 1966–83. *J. Anim. Ecol.,* 53:327–350.

Emlen, J. M. 1973. *Ecology: an evolutionary approach.* Reading, Mass.: Addison-Wesley.

Enomoto, Y. 1956. On the occurrence and the food of *Noctiluca scintillans* (Macartney) in the waters adjacent to the west coast of Kyushu, with special reference to the possibility of the damage caused to the fish eggs by that plankton. *Bull. Jap. Soc. Sci. Fish.,* 22:82–88.

Fabre-Domergue, M. 1900. Etude sur la rôle et les procedes de la pisiculture marine. *Bull. Mar. Marchande.* Paris, 1900.

Fabre-Domergue, M., and E. Bietrix. 1905. Developpement de la sole. Introduction à l'Etude de la pisiculture marine. Paris, 1905.

Feller, W. 1957. *An introduction to probability theory and its applications.* New York: Wiley.

Food and Agriculture Organization of the United Nations. 1978. Some scientific problems of multispecies fisheries. Report of the Expert Consultation on Management of Multispecies Fisheries. FAO Fish. Tech. Paper No. 181.

Fox, W. W. 1970. An exponential yield model for optimizing exploited fish populations. *Trans. Am. Fish. Soc.,* 99:80–88.

Fransz, H. G., and W. W. C. Gieskes. 1984. The unbalance of phytoplankton and copepods in the North Sea. *Rapp. P.-v. Réun. Cons. int. Explor. Mer,* 183:218–225.

Fridgeirsson, E. 1984. Cod larvae sampling with a large pump off SW-Ireland. In E. Dahl, D. S. Danielssen, E. Moksness, and P. Solemdal, eds., *The propagation of cod, Gadus morhua L.* Flødevigen rapportser 1, 1984:317–333.

Friedrich, H. 1983. Simulation of the thermal stratification at the FLEX Central Station with a one-dimensional integral model. In J. Sundermann and W. Lenz, eds., *North Sea dynamics.* Berlin: Springer-Verlag.

Fye, P. M. 1971. Some quotes. *Oceanus,* 16:1.

Garcia, S. 1983. The stock-recruitment relationship in shrimps: Reality or artefacts and misinterpretations? *Oceanogr. Trop.,* 18:25–48.

Garrod, D. J. 1973. The variation of replacement and survival in some fish stocks. *Rapp. P.-v. Réun Cons. int. Explor. Mer,* 164:43–56.

———— 1977. The North Atlantic cod. In J. A. Gulland, ed., *Fish population dynamics.* New York: Wiley.

———— 1982. Stock and recruitment—again. Fish. Res. Tech. Rep., MAFF Direct. Fish. Res., Lowestoft (68).

Garrod, D. J., and B. J. Knights. 1979. Fish stocks: their life-history characteristics and response to exploitation. *Symp. Zool. Soc. Lond.,* 44:361–382.

Gerritsen, J., and J. R. Strickler. 1977. Encounter probabilities and community structure in zooplankton: a mathematical model. *J. Fish. Res. Board Can.,* 34:73–82.

Gieskes, W. W. C., and W. Kraay. 1984. State-of-the-art in the measurement of primary production. In M. J. R. Fasham, ed., *Flows of energy and materials in marine ecosystems.* NATO Conference Series IV, Marine Science; V. 13. New York: Plenum Press.

Glantz, M. H., and J. D. Thompson, eds. 1981. *Resource management and environmental uncertainty.* New York: Wiley.

Goldman, J. C. 1984. Oceanic nutrient cycles. In M. J. R. Fasham, ed, *Flows of energy and materials in marine ecosystems.* NATO Conference Series IV, Marine Science; V. 13. New York: Plenum Press.

Graham, M. 1924. The annual cycle in the life of the mature cod in the North Sea. *Min. of Agric. and Fisheries. Fishery Investigations II,* vol. 6.

———— 1935. Modern theory of exploiting a fishery, and application of North Sea trawling. *J. Cons. int. Explor. Mer,* 10:263–274.

Grave, H. 1981. Food and feeding of mackerel larvae and early juveniles in the North Sea. *Rapp. P.-v. Réun. Cons. int. Explor. Mer,* 178:454–459.

Gross, M. R., and R. C. Sargent. 1985. The evolution of male and female parental care in fishes. *Amer. Zool.,* 25:807–822.

Gulland, J. A. 1953. Correlations on fisheries hydrography. *J. Cons. int. Explor. Mer,* 18:351–353.

———— 1977. The stability of fish stocks. *J. Cons. int. Explor. Mer,* 37:199–204.

———— 1982. Why do fish numbers vary? *J. Theor. Biol.,* 97:69–75.

Gulland, J. A., and L. K. Boerema. 1973. Scientific advice on catch levels. *Fish. Bull. U.S.,* 71:325–335.

Gulland, J. A., and S. Garcia. 1984. Observed patterns in multispecies fisheries. In R. M. May, ed., *Exploitation of marine communities.* Berlin: Springer-Verlag.

Gumbel, E. J. 1959. *Statistics of extremes.* New York: Columbia University Press.

Hammack, J. M. 1969. Toward an economic evaluation of a fugitive recreational resource: Waterfowl. Unpubl. thesis, University of Washington.

Hanavan, M. G., and B. E. Skud. 1954. Intertidal spawning of pink salmon. *Fish. Bull.,* U.S.

Hardin, G. 1968. The tragedy of the commons. *Science,* 162:1243–248.

Harding, D., J. H. Nichols, and D. S. Tungate. 1978. The spawning of plaice (*Pleuronectes platessa* L.) in the southern North Sea and English Channel. *Rapp. P.-v. Réun. Cons. int. Explor. Mer,* 172:102–113.

Harding, D., and J. W. Talbot. 1973. Recent studies on the eggs and larvae of the plaice (*Pleuronectes platessa* L.) in the southern bight. *Rapp. P.-v. Réun. Cons. int. Explor. Mer,* 164:261–269.

Harris, J. G. K. 1975. The effect of density-dependent mortality on the shape of the stock and recruitment curve. *J. Cons. int. Explor. Mer,* 36:144–149.

Hattori, S. 1962. Predatory activity of *Noctiluca* on achovy eggs. *Bull. Tokai Reg. Fish. Res. Lab.,* 9:211–220.

Haury, L. R., J. A. McGowan, and P. H. Wiebe. 1978. Patterns and processes in the time-space scales of plankton distributions. In J. H. Steele, ed., *Spatial pattern in plankton communities.* NATO Conference Series 4, Marine Sciences, vol. 3. New York: Plenum Press.

Hayashi, S. 1961. Fishery biology of the Japanese anchovy, *Engraulis japonica* (Houttuyn). *Bull. Tokai Reg. Fish. Res. Lab.,* 31:145–268.

Hempel, G. 1978a. Symposium on North Sea fish stocks—recent changes and their causes. *Rapp. P.-v. Réun. Cons. int. Explor. Mer,* 172:5–9.

——— 1978b. North Sea fisheries and fish stocks—a review of recent changes. *Rapp. P.-v. Réun. Cons. int. Explor. Mer,* 173:145–167.

——— 1979. *Early life history of marine fish.* Seattle: Washington Sea Grant Program.

Hempel, G., and J. H. S. Blaxter. 1963. On the condition of herring larvae. *Rapp. P.-v. Réun Cons. int. Explor. Mer,* 154:35–40.

Hempel, I., and G. Hempel. 1971. An estimate of mortality of eggs of North Sea herring (*Clupea harengus* L.). *Rapp. P.-v. Réun. Cons. int. Explor. Mer,* 160:24–26.

Hennemuth, R. C., J. E. Palmer, and B. E. Brown. 1982. A statistical description of recruitment in eighteen selected fish stocks. *J. Northw. Atl. Fish. Sci.,* 1:101–111.

Hester, F. J. 1964. Effects of food supply on fecundity in the female guppy *Lebistes reticulatus* (Peters). *J. Fish. Res. Board Can.,* 21:757–764.

Hislop, J. R. G. 1975. The breeding and growth of whiting, *Merlangius merlangus,* in captivity. *J. Cons. int. Explor. Mer,* 36:119–127.

Hislop, J. R. G., A. P. Robb, and J. A. Gauld. 1978. Observations on effects of feeding level on growth and reproduction in haddock, *Melanogrammus aeglefinus* (L.), in captivity. *J. Fish. Biol.,* 13:85–98.

Hislop, J. R. G., and A. M. Shanks. 1981. Recent investigations on the reproductive biology of the haddock, *Melanogrammus aeglefinus,* of the northern North Sea and the effects on fecundity of infection with the copepod parasite *Lernaeocera branchialis. J. Cons. int. Explor. Mer,* 39:244–251.

Hjort, J. 1914. Fluctuations in the great fisheries of northern Europe viewed in the light of biological research. *Rapp. P.-v. Réun. Cons. int. Explor. Mer*, 20:1–228.

—— 1926. Fluctuations in the year classes of important food fishes. *J. Cons. int. Explor. Mer*, 1:5–38.

—— 1932. Remarks on the fluctuations in number and growth in marine populations. *Rapp. P.-v. Réun. Cons. int. Explor. Mer*, 80:1–8.

—— 1938. Studies of growth in the north-eastern area. *Rapp. P.-v. Réun. Cons. int. Explor. Mer*, 108:1–8.

Holden, M. J. 1972. Variations in year class strengths of cod in the English North Sea fisheries from 1954 to 1967 and their relation to sea temperature. *Annls. Biol.*, Copenh., 27:86.

—— 1978. Long-term changes in landings of fish from the North Sea. *Rapp. P.-v. Réun. Cons. int. Explor. Mer*, 172:11–26.

Holling, C. S. 1965. The functional response of invertebrate predators to prey density and its role in mimicry and population regulation. *Mem. Entomol. Soc. Canada*, 45:1–60.

Holt, S. J., and L. M. Talbot. 1978. New principles for the conservation of wild living resources. *J. Wildl. Mgmt., Wildlife Monograph*, 59:1–33.

Houde, E. D. 1972. Development and early life history of the northern sennet, *Sphyraena borealis* DeKay (Pisces: Sphyraenidae), reared in the laboratory. *Fish. Bull.*, U.S., 70:185–195.

—— 1973. Some recent advances and unsolved problems in the culture of marine fish larvae. *Proc. Wld. Maricult. Soc.*, 3:83–112.

—— 1974. Effects of temperature and delayed feeding on growth and survival of larvae of three species of subtropical marine fishes. *Mar. Biol.*, 26:271–285.

—— 1975. Effects of stocking density and food density on survival, growth, and yield of laboratory-reared larvae of sea bream *Archosargus rhomboidalis* (L.) (Sparidae). *J. Fish. Biol.*, 7:115–127.

—— 1977. Food concentrations and stocking density effects on survival and growth of laboratory-reared larvae of bay anchovy *Anchoa mitchilli* and lined sole *Achirus lineatus. Mar. Biol.* 43:333–341.

—— 1978. Critical food concentrations for larvae of three species of subtropical marine fishes. *Bull. Mar. Sci.*, 28:395–411.

Houde, E. D., and J. D. Alpern Lovdal. 1984. Seasonality of occurrence, foods, and food preferences of ichthyoplankton in Biscayne Bay, Florida. *Estuarine, Coastal, and Shelf Science*, 18:403–419.

Houde, E. D., and B. J. Palko. 1970. Laboratory rearing of the clupeid fish *Harengula pensacolae* from fertilized eggs. *Mar. Biol.*, 5:354–358.

Houde, E. D., and R. C. Schekter. 1978. Simulated food patches and survival of larval bay anchovy, *Anchoa mitchilli*, and sea bream, *Archosargus rhomboidalis. Fish. Bull.*, U.S., 76:483–487.

—— 1980. Feeding by marine fish larvae: Developmental and functional responses. *Env. Biol. Fish.*, 5:315–334.

Hubold, G. 1978. Variations in growth rate and maturity of herring in the northern North Sea in the years 1955–1973. *Rapp. P.-v. Réun. Cons. int. Explor. Mer*, 172:154–163.

Hunter, J. R. 1972. Swimming and feeding behavior of larval anchovy, *Engraulis mordax*. *Fish. Bull.*, U.S., 70:821–838.

——— 1976. Culture and growth of northern anchovy *Engraulis mordax* larvae. *Fish Bull.*, U.S., 74:81–88.

——— 1977. Behavior and survival of northern anchovy *Engraulis mordax* larvae. *Calif. Coop. Oceanic Fish. Invest. Rep.*, 19:138–146.

——— 1980. The feeding behavior and ecology of marine fish larvae. In J. E. Bardach, J. J. Magnuson, R. C. May, and J. M. Reenhart, eds., *Fish behavior and its use in the capture and culture of fishes*. ICLARM Conf. Proc. 5., Manila.

——— 1981. Feeding ecology and predation of marine fish larvae. In R. Lasker, ed., *Marine fish larvae*. Seattle: Washington Sea Grant Program.

——— 1982. Predation and recruitment. In Rothschild and Rooth, 1982.

——— 1984. Inferences regarding predation on the early life stages of cod and other fishes. In E. Dahl, D. S. Danielssen, E. Moksness, and P. Solemdal, eds., *The propagation of cod, Gadus morhua L.* Flødevigen rapportser 1, 1984:533–562.

Hunter, J. R., and S. R. Goldberg. 1980. Spawning incidence and batch fecundity in northern anchovy, *Engraulis mordax*. *Fish. Bull.*, U.S., 77:641–652.

Hunter, J. R., and C. A. Kimbrell. 1980. Egg cannibalism in the northern anchovy, *Engraulis mordax*. *Fish. Bull.*, U.S., 78:811–816.

Hunter, J. R., and R. Leong. 1980. The spawning energetics of female northern anchovy, *Engraulis mordax*. *Fish. Bull.*, U.S., 79:215–230.

Hunter, J. R., and G. L. Thomas. 1974. Effect of prey distribution and density on the searching and feeding behavior of larval anchovy *Engraulis mordax* Girard. In Blaxter, 1974.

Husby, D. N., and C. S. Nelson. 1982. Turbulence and vertical stability in the California current. *Calif. Coop. Oceanic Fish. Invest. Progr. Rep.*, 23:113–129.

Hutchinson, G. E. 1978. *An introduction to population ecology*. New Haven: Yale University Press.

Iles, T. D. 1964. The duration of maturation stages in herring. *J. Cons. int. Explor. Mer*, 29:166–188.

——— 1965. Factors determining or limiting the physiological reaction of herring to environmental change. International Commission for the Northwest Atlantic Fisheries. Special Publ. No. 6:735–741.

——— 1967. Growth studies of North Sea herring, I. The second year's growth (I-group) of East Anglian herring, 1939–63. *J. Cons. int. Explor. Mer*, 31:56–76.

——— 1968. Growth studies of North Sea herring, II. O-group growth of East Anglian herring. *J. Cons. int. Explor. Mer*, 32:98–116.

——— 1973. Interaction of environment and parent stock size in determining recruitment in the Pacific sardine as revealed by analysis of density-dependent O-group growth. *Rapp. P.-v. Réun. Cons. int. Explor. Mer*, 164:228–240.

——— 1984. Allocation of resources to gonad and soma in Atlantic herring

Clupea harengus L. In G. W. Potts and R. J. Wootton, eds., *Fish reproduction: strategies and tactics.* London: Academic Press.

Iles, T. D., and M. Sinclair. 1982. Atlantic herring: Stock discreteness and abundance. *Science,* 215:627–633.

Iles, T. D., and R. J. Wood. 1965. The fat/water relationship in North Sea herring (*Clupea harengus*) and its possible significance. *J. Mar. Biol. Ass., U.K.,* 45:353–366.

Jakobsson, J. 1985. Monitoring and management of the Northeast Atlantic herring stocks. *Can. J. Fish. Aquat. Sci.,* 42:207–221.

Jakobsson, J., and O. Halldórsson. 1984. Changes in biological parameters in the icelandic summer spawning herring. ICES C.M. 1984/H:43.

Jensen, A. J. C. 1927. On the influence of the quantity of spawning herrings upon the stock of the following years. *J. Cons. int. Explor. Mer,* 2:44–49.

Johnson, W. E. 1965. On mechanisms of self-regulation of population abundance in *Oncorhynchus nerka. Mitt. Intern. Ver. Limnol.,* Stuttgart, 13:66–87.

Jones, R. 1973. Density dependent regulation of the numbers of cod and haddock. *Rapp. P.-v. Réun. Cons. int. Explor. Mer,* 164:156–173.

———— 1983. The decline in herring and mackerel and the associated increase in other species in the North Sea. In G. D. Sharp and J. Csirke, eds., *Proceedings of the Expert Consultation to Examine Changes in Abundance and Species Composition of Neritic Fish Resources.* FAO Fisheries Report No. 291:507–520.

———— 1984. Some observations on energy transfer through the North Sea and Georges Bank food webs. *Rapp. P.-v. Réun. Cons. int. Explor. Mer,* 183:204–217.

Jones, R., and W. B. Hall. 1973. A simulation model for studying the population dynamics of some fish species. In M. S. Bartlett and R. W. Hiorns, eds., *The mathematical theory of the dynamics of biological populations.* London and New York: Academic Press.

———— 1974. Some observations on the population dynamics of the larval stage in the common gadoids. In Blaxter, 1974.

Junge, C. O. 1966. Depensatory process based on the concept of hunger. *J. Fish. Res. Board Can.,* 23:689–699.

Kawai, T., and K. Isibasi. 1983. Change in abundance and species composition of neritic pelagic fish stocks in connection with larval mortality caused by cannibalism and predatory loss by carnivorous plankton. In G. D. Sharp, and J. Csirke, eds., *Proceedings of the Expert Consultation to Examine Changes in Abundance and Species Composition of Neritic Fish resources.* FAO Fisheries Report No. 291:1081–111.

Koopman, B. O. 1956. The theory of search. I. Kinematic bases. *Operations Research,* 4:324–531.

Lack, D. 1954. *The natural regulation of animal numbers.* London: Oxford University Press.

———— 1966. *Population studies of birds.* Oxford: Clarendon Press.

Larkin, P. A. 1977. An epitaph for the concept of maximum sustained yield. *Trans. Am. Fish. Soc.* 106:1–11.

Larkin, P. A., R. F. Raleigh, and N. J. Wilimovsky. 1964. Some alternative premises for constructing theoretical reproduction curves. *J. Fish. Res. Board Can.*, 21:477–484.

Lasker, R. 1975. Field criteria for survival of anchovy larvae: the relation between inshore chlorophyll maximum layers and successful first feeding. *Fish. Bull.*, U.S., 73:453–462.

———— 1985. What limits clupeoid production? *Can. J. Fish. Aquat. Sci.*, 42:31–38.

Lasker, R., and A. MacCall. 1983. New ideas on the fluctuations of the clupeoid stocks off California. *Proceedings of the Joint Ocean. Assembly, 1982 — General Symposia.* Can. Nat. Comm., Ottawa.

Lasker, R., and K. Sherman. 1981. The early life history of fish: recent studies. *Rapp. P.-v. Réun. Cons. int. Explor. Mer*, 178:1–607.

Lasker, R., H. W. Feder, G. H. Theilacker, and R. C. May. 1970. Feeding growth and survival of *Engraulis mordax* larvae reared in the laboratory. *Mar. Biol.*, 5:345–353.

Last, J. M. 1978a. The food of four species of pleuronectiform larvae in the eastern English Channel and southern Northern Sea. *Mar. Biol.*, 45:359–368.

———— 1978b. The food of three species of gadoid larvae in the eastern English Channel and southern North Sea. *Mar. Biol.*, 48:377–386.

Laurence, G. C. 1974. Growth and survival of haddock (*Melanogrammus aeglefinus*) larvae in relation to planktonic prey concentration. *J. Fish. Res. Board Can.* 31:1415–19.

———— 1982. Nutrition and trophodynamics of larval fish—review, concepts, strategic recommendations, and opinions. In Rothschild and Rooth, 1982.

Lebour, M. V. 1920. The food of young fish. No. 3 (1919). *J. Mar. Biol. Ass.*, U.K., 12:261–324.

Ledbetter, M. 1979. Langmuir circulations and plankton patchiness. *Ecol. Mod.*, 7:289–310.

Leggett, W. C. 1977. Density dependence, density independence, and recruitment in the American shad (*Alosa sapidissima*) population of the Connecticut River. In W. Van Winkle, ed., *Assessing the effects of power-plant-induced mortality on fish poulations.* New York: Pergamon Press.

Lewin, R. 1983. Santa Rosalia was a goat. *Science*, 221:636–639.

Lockwood, S. J. 1980. Density-dependent mortality in 0-group plaice (*Pleuronectes platessa* L.) populations. *J. Cons. int. Explor. Mer*, 39:148–153.

Longwell, A. C., and J. B. Hughes. 1981. Cytologic, cytogenetic, and embryologic state of Atlantic mackerel eggs from surface waters of the New York bight in relation to pollution. *Rapp. P.-v. Réun. Cons. int. Explor. Mer*, 178:76–78.

Lotka, A. J. 1925. *Elements of physical biology.* Baltimore: Williams and Wilkins.

———— 1956. *Elements of mathematical biology.* New York: Dover.

Ludwig, D., and C. Walters. 1982. Measurement errors and uncertainty in parameters estimates for stock and recruitment. *Can. J. Fish. Aquat. Sci.*, 38:711–720.

Lyagina, T. N. 1975. Connection of egg weight with biological indices of female roach, *Rutilus rutilus*, with different abundance of food. *J. Ichthyology*, 15:584–594.

MacCall, A. D. 1979. Population estimates for the waning years of the Pacific sardine fishery. *Calif. Coop. Oceanic Fish. Invest. Rep.*, 20:72–82.

―――― 1980a. Population models for the northern anchovy *(Engraulis mordax)*. *Rapp. P.-v. Réun. Cons. int. Explor. Mer*, 177:292–306.

―――― 1980b. The consequences of cannibalism in the stock-recruitment relationship of planktivorous pelagic fishes such as *Engraulis*. Intergovernmental Oceanography Commission Workshop Report No. 28:201–220. FAO.

―――― 1983. Variability of pelagic fish stocks off California. In G. D. Sharp and J. Csirke, eds., In *Proceedings of the Expert Consultation to Examine Changes in Abundance and Species Composition of Neritic Fish Resources*. FAO Fisheries Report No. 291:101–112.

Mandlebrot, B. B. 1977. *Fractals: form, chance, and dimension*. San Francisco: Freeman.

Marak, R. R. 1974. Food and feeding of larval redfish in the Gulf of Maine. In Blaxter, 1974.

Marr, J. C. 1956. The "critical period" in the early life history of marine fishes. *J. Cons. int. Explor. Mer*, 21:160–170.

―――― 1960. The causes of major variations in the catch of the Pacific sardine *Sardinops caerulea* (Girard). *Proc. World Sci. Meetings Biology of Sardines and Related Species*, FAO 3:667–679.

―――― 1963. A model of the population biology of the Pacific sardine *Sardinops caerulea*. *Rapp. P.-v. Réun. Cons. int. Explor. Mer*, 154:270–278.

May, R. C. 1974. Larval mortality in marine fishes and the critical period concept. In Blaxter, 1974.

May, R. M. 1973. *Stability and complexity in model ecosystems*. Princeton, N. J.: Princeton University Press.

McCarthy, J. 1984. Measuring oceanic primary production. In *Global ocean flux study*. Proceedings of a workshop September 10–14, 1984, National Academy of Sciences, Woods Hole Study Center, Woods Hole, Mass.

Mercer, M. C., ed. 1982. Multispecies approaches to fisheries management advice. *Can. Spec. Publ. Fish. Aquat. Sci.*, 59:1–169.

Merthot, R. D., Jr. 1983. Seasonal variation in survival of larval Northern Anchovy, *Engraulis mordax*, estimated from the age distribution of juveniles. *Fish. Bull.* 81:741–750.

Moffatt, N. M. 1981. Survival and growth of northern anchovy larvae on low zooplankton densities as affected by the presence of a *Chlorella* bloom. *Rapp. P.-v. Réun. Cons. int. Explor. Mer*, 178:475–480.

Moser, H. G. 1981. Morphological and functional aspects of marine fish larvae. In R. Lasker, ed., *Marine fish larvae*. Seattle: Washington Sea Grant Program.

Murphy, G. I. 1966. Population biology of the Pacific sardine *(Sardinops caerulea)*. *Proceedings, Calif. Acad. Sci.*, 34:1–84.

―――― 1973. *Clupeoid fishes under exploitation with special reference to the Peruvian anchovy*. Tech. Rep. No. 30, Hawaii Inst. of Marine Biology.

―――― 1977. Clupeoids. In J. A. Gulland, ed., *Fish population dynamics*. New York: Wiley.

Myers, J. H., and C. J. Krebs. 1974. Population cycles in rodents. *Scient. Am.*, 230:38–46.

Neave, F. 1952. "Even-year" and "odd-year" pink salmon populations. *Trans. Royal Soc. Canada*, V, 46:55–70.

Nikol'skii, G. V. 1962. On some adaptations to the regulation of population density in fish species with different types of stock structure. In E. D. LeCren and M. W. Holdgate, ed., *The exploitation of natural animal populations*. Oxford: Blackwell Scientific Publications.

———— 1963. *The ecology of fishes*. Trans. from the Russian by L. Birkett. New York: Academic Press.

Nisbet, R. M., and W. S. C. Gurney. 1982. Modelling fluctuating populations. New York: Wiley.

Nordeng, H., and T. Bratland. 1971. Feeding of plaice (*Pleuronectes platessa* L.) and cod (*Gadus morhua* L.) larvae. *J. Cons. int. Explor. Mer*, 34:51–57.

O'Boyle, R. N., M. Sinclair, R. J. Conover, K. H. Mann, and A. C. Kohler. 1984. Temporal and spatial distribution of ichthyoplankton communities of the Scotian Shelf in relation to biological, hydrological, and physiographic features. *Rapp. P.-v. Réun. Cons. int. Explor. Mer*, 183:27–40.

O'Connell, C. P. 1980. Percentage of starving northern anchovy, *Engraulis mordax*, larvae in the sea as estimated by histological methods. *Fish. Bull., U.S.*, 78:475–489.

O'Connell, C. P., and L. P. Raymond. 1970. The effect of food density on survival and growth of early post yolk-sac larvae of the northern anchovy (*Engraulis mordax* Girard) in the laboratory. *J. Exp. Mar. Biol. Ecol.*, 5:187–197.

Olafsen, J. A. 1984. Ingestion of bacteria by cod (*Gadus morhua* L.) larvae. In E. Dahl, D. S. Danielssen, E. Moksness, and P. Solemdal, eds., *The propagation of cod*, Gadus morhua L. Flødevigen rapportser 1, 1984:627–639.

Paffenhoffer, G. A. 1976. Feeding, growth, and food conversion of the marine planktonic copepod *Calanus helgolandicus*. *Limnol. Oceanogr.*, 21:39–50.

Paulik, G. J. 1973. Studies of the possible form of the stock-recruitment curve. *Rapp. P.-v. Réun. Cons. int. Explor. Mer*, 164:303–315.

Paulik, G. J., and J. W. Greenough, Jr. 1966. Management analysis for a salmon resource system. In K. E. F. Watt, ed., *Systems analysis in ecology*. New York: Academic Press.

Pauly, D. 1980. On the interrelationships between natural mortality, growth parameters, and mean environmental temperature in 175 fish stocks. *J. Cons. int. Explor. Mer*, 39:175–192.

Pauly, D., and G. I. Murphy, eds. 1982. *Theory and management of tropical fisheries*. ICLARM Conf. Proc. 9, Manila.

Pedersen, T. 1984. Variation in peak spawning of arcto-Norwegian cod (*Gadus morhua* L.) during the time period 1929–1982 based on indices estimated from fishery statistics. In E. Dahl, D. S. Danielssen, E. Moksness, and P. Solemdal, eds., *The propagation of cod*, Flødevigen rapportser 1, 1984:301–316.

Pella, J. J. and P. K. Tomlinson. 1969. A generalized stock production model. *Bull. Inter-Am. Trop. Tuna Comm.* 13:421–458.

Perkins, R. J., and M. D. Dahlberg. 1971. Fat cycles and condition factors of Altamaha River shads. *Ecology,* 52:358–362.

Peterman, R. M.; W. C. Clark; and C. S. Holling. 1979. The dynamics of resilience: Shifting stability domains in fish and insect systems. In R. M. Anderson, B. D. Turner, and L. R. Taylor, eds., *Population dynamics.* Oxford: Blackwell Scientific Publications.

Pielou, E. C. 1977. *Mathematical ecology.* New York: Wiley.

Platt, T., M. Lewis, and R. Geider. 1984. Thermodynamics of the pelagic ecosystem: Elementary closure conditions for biological production in the open ocean. In M. J. R. Fasham, ed., *Flows of energy and materials in marine ecosystems.* NATO Conference Series IV, Marine Sciences; V. 13. New York: Plenum Press.

Pomeroy, L. R. 1974. The ocean's food web, a changing paradigm. *BioScience,* 24:499–504.

Pommeranz, T. 1974. Resistance of plaice eggs to mechanical stress and light. In Blaxter, 1974.

———1981. Observations on the predation of herring (*Clupea harengus* L.) and sprat (*Sprattus sprattus* L.) on fish eggs and larvae in the southern North Sea. *Rapp. P.-v. Réun. Cons. int. Explor. Mer,* 178:402–404.

Radovich, J. 1981. The collapse of the California sardine fishery—what have we learned? In M. H. Glantz and J. D. Thompson, eds., *Resource management and environmental uncertainty.* New York: Wiley.

Raitt, D. F. S. 1939. The rate of mortality of the haddock in the North Sea stock, 1919–1938. *Rapp. P.-v. Réun. Cons. int. Explor. Mer,* 110:65–79.

Rashevsky, N. 1959. Some remarks on the mathematical theory of nutrition of fishes. *Bull. Math. Biophysics,* 21:161–183.

Rauck, G., and J. J. Zijlstra. 1978. On the nursery-aspects of the Wadden Sea for some commercial fish species and possible long-term changes. *Rapp. P.-v. Réun. Cons. int. Explor. Mer,* 172:266–275.

Report of the fourth session of the panel of experts on stock assessment on Peruvian anchoveta. 1974. Bol. Inst. Mar, Peru-Callao, 2:605–719.

Ricker, W. E. 1954. Stock and recruitment. *J. Fish. Res. Board Can.,* 11:559–623.

———1958. Handbook of computations for biological statistics of fish populations. *Bull. Fish. Res. Board Can.,* 119:1–300.

———1962. Regulation of the abundance of pink salmon populations. In N. J. Wilimovsky, ed. *Symposium on Pink Salmon.* Vancouver: University of British Columbia Press.

———1973. Critical statistics from two reproduction curves. *Rapp. P.-v. Réun. Cons. int. Explor. Mer,* 164:333–340.

———1975. Computation and interpretation of biological statistics of fish populations. *Bull. Fish. Res. Board Can.,* 119:382.

———1979. Growth rates and models. In W. S. Hoar and D. J. Randall, eds., *Fish physiology,* vol. 8. New York: Academic Press.

Riley, G. A. 1963. Theory of food chain relations in the ocean. In M. N. Hill, ed., *The Sea,* vol 2. New York: Wiley.

Rollefsen, G. 1930. Observations on cod eggs. *Rapp. P.-v. Réun. Cons. Perm. int. Explor. Mer,* 65:31–34.

———1932. The susceptibility of cod eggs to external influences. *J. Cons. int. Explor. Mer*, 7:367–373.

Rosenthal, H., and G. Hempel. 1970. Experimental studies in feeding and food requirements of herring larvae (*Clupea harengus* L.). In J. H. Steele, ed., *Marine food chains*. Berkeley: University of California Press.

Rothschild, B. J. 1961. Production and survival of eggs of the American smelt, *Osmerus mordax* (Mitchill), in Maine. *Trans. Am. Fish. Soc.*, 90:42–48.

———1967. Competition for gear in a multiple species fishery. *J. Cons. int. Explor. Mer*, 31:102–110.

———1971. A systems view of fishery management with some notes on the tuna fisheries. FAO Fish. Techn. Paper No. 106.

———1977. Fishing effort. In J. Gulland, ed., *Fish population dynamics*. London: Wiley.

———1981. More food from the sea? *BioScience*, 31:216–222.

———1983a. Achievement of goals in fisheries management. In B. J. Rothschild, ed., *Global fisheries: perspectives for the 1980's*. New York: Springer-Verlag.

———1983b. On the allocation of fisheries stock. In J. W. Reintjes, ed., *Improving multiple use of coastal and marine resources*. American Fisheries Society. Laurence, Kansas: Allen Press.

Rothschild, B. J., and S. L. Brunenmeister. 1984. Temporal patterns in recruitment data. ICES C.M. 1984/G:14.

Rothschild, B. J., and D. G. Heimbuch. 1983. Managing variable fishery stocks in an uncertain environment: the design of fishery management systems. In G. D. Sharp and J. Csirke, eds., *Proceedings of the Expert Consultation to Examine Changes in Abundance and Species Composition of Neritic Fish Resources*. FAO Fisheries Report No. 291:1141–159.

Rothschild, B. J., E. D. Houde, and R. Lasker. 1982. Causes of fish stock fluctuation: problem setting and perspectives. In Rothschild and Rooth, 1982.

Rothschild, B. J., and A. J. Mullen. 1985. The information content of stock-and-recruitment data and its non-parametric classification. *J. Cons. int. Explor. Mer*, 42:116–124.

Rothschild, B. J., and C. Rooth, eds. 1982. *Fish ecology III. A foundation for REX. A recruitment experiment*. Miami: University of Miami Technical Report No. 82008.

Rothschild, B. J., and A. Suda. 1977. Population dynamics of tuna. In J. Gulland, ed., *Fish population dynamics*. London: Wiley.

Rounsefell, G. A., and G. B. Kelez. 1938. The salmon and salmon fisheries of Swiftsure Bank, Puget Sound, and the Fraser River. *Fish. Bull., U.S.*, 49:693–823.

Runnström, S. 1941. Racial analysis of the herring in Norwegian waters. *FiskDir. Skr. Ser. HavUnders*, 6(7):1–110.

Russell, E. S. 1931. Some theoretical considerations on the "overfishing" problem. *J. Cons. int. Explor. Mer*, 6:3–20.

Saksena, V. P., and E. D. Houde. 1972. Effect of food level on the growth and survival of laboratory-reared larvae of bay anchovy (*Anchoa mitchilli* Valenciennes) and scaled sardine (*Harengula pensacolae* Good and Bean). *J. Exp. Mar. Biol. Ecol.*, 8:249–258.

Saville, A. 1959. The planktonic stages of the haddock in Scottish waters. *Mar. Res.*, 3:1–23.

Schaefer, M. B. 1954. Some aspects of the dynamics of populations important to the management of the commercial marine fisheries. *Bull. Inter-Am. Trop. Tuna Comm.*, 1:27–56.

——— 1957. A study of the dynamics of fishery for yellowfin tuna in the eastern tropical Pacific Ocean. *Bull. Inter-Am. Trop. Tuna Comm.*, 2:247–285.

Schaefer, M. B., and R. J. H. Beverton. 1963. Fishery dynamics—their analysis and interpretation. In M. N. Hill, ed., *The Sea*, vol. 2. New York: Wiley.

Schlee, S. 1973. *A history of oceanography: the edge of an unfamiliar world*. London: Robert Hale.

Scott, D. P. 1962. Effect of food quantity on fecundity of rainbow trout *Salmo gairdneri*. *J. Fish. Res. Board Can.*, 19:715–731.

Scura, E. D., and C. W. Jerde. 1977. Various species of phytoplankton as food for larval northern anchovy *Engraulis mordax*, and relative nutritional value of the dinoflagellate *Gymnodinium splendens* and *Gonyaulax polyedra*. *Fish. Bull.*, U.S., 75:577–583.

Sette, O. E. 1943. Biology of the Atlantic mackerel *(Scomber scombrus)* of North America. Part I. Early life history, including growth, drift and mortality of the egg and larval populations. *Fish. Bull.*, U.S., 38:149–237.

Shannon, C. E., and W. Weaver. 1959. *The mathematical theory of communication*. Urbana: University of Illinois Press.

Shelbourne, J. E. 1957. The feeding and condition of plaice larvae in good and bad plankton patches. *J. Mar. Biol. Ass.*, U.K. 36:539–552.

Sheldon, R. W., A. Prakash, and W. H. Sutcliffe, Jr. 1972. The size distribution of particles in the ocean. *Limnol. Oceanogr.*, 17:327–340.

Shelton, P. A., and M. J. Armstrong. 1983. Variations in the parent stock and recruitment of pilchard and anchovy populations in the southern Benguela system. In G. D. Sharp, and J. Csirke, eds., *Proceedings of the Expert Consultation to Examine Changes in Abundance and Species Composition of Neritic Fish Resources*. FAO Fisheries Report No. 291:1113–132.

Shepherd, J. G. 1982. A versatile new stock-recruitment relationship for fisheries, and the construction of sustainable yield curves. *J. Cons. int. Explor. Mer*, 40:67–75.

Shepherd, J. G., and D. H. Cushing. 1980. A mechanism for density-dependent survival of larval fish as the basis of a stock-recruitment relationship. *J. Cons. int. Explor. Mer*, 39:160–167.

Shepherd, J. G., J. G. Pope, and R. D. Cousens. 1984. Variations in fish stocks and hypotheses concerning their links with climate. *Rapp. P.-v. Réun. Cons. int. Explor. Mer*, 185:255–267.

Silliman, R. P. 1975. Experimental exploitation of competing fish populations. *Fish. Bull.*, U.S., 73:872–888.

Silvert, W. 1981. Principles of ecosystem modelling. In A. R. Longhurst, ed., *Analysis of marine ecosystems*. London: Academic Press.

Sinclair, M., and M. J. Tremblay. 1984. Timing of spawning of Atlantic herring *(Clupea harengus harengus)* populations and the match-mismatch theory. *Can. J. Fish. Aquat. Sci.*, 41:1055–65.

Sissenwine, M. P. 1984. Why do fish populations vary? In R. M. May, ed., *Exploitation of marine communities.* Berlin: Springer-Verlag.

Sissenwine, M. P., B. E. Brown, J. E. Palmer, R. J. Essig, and W. Smith. 1982. Empirical examination of population interactions for the fishery resources off the northeastern U.S.A. In Mercer, 1982.

Skellam, J. G. 1971. Some philosophical aspects of mathematical modeling in empirical science with special reference to ecology. In J. N. R. Jeffers, ed., *Mathematical models in ecology.* Twelfth Symposium of British Ecological Society. Grange-over-Sands, Lancashire. Oxford: Blackwell Scientific Publications.

Skud, B. E. 1982. Dominance in fishes: the relation between environment and abundance. *Science,* 216:144–149.

Smith, P. E. 1973. The mortality and dispersal of sardine eggs and larvae. *Rapp. P.-v. Réun. Cons. int. Explor. Mer,* 164:282–292.

Soleim, P. A. 1942. Arsaker til rike og fattige arganger av sild. *FiskDir. Skr. Ser. HavUnders,* 7:1–39.

Soutar, A., and J. D. Isaacs. 1969. History of fish populations inferred from fish scales in anaerobic sediments off California. *Calif. Coop. Oceanic Fish Invest. Rep.,* 13:63–70.

———— 1974. Abundance of pelagic fish during the nineteenth and twentieth centuries as recorded in anaerobic sediment off California. *Fish. Bull., U.S.,* 72:257–273.

Southward, A. J., and N. Demir. 1974. Seasonal changes in dimensions and viability of the developing eggs of the cornish pilchard (*Sardina pilchardus* Walbaum) off Plymouth. In Blaxter, 1974.

Stacey, N. E. 1984. Control of the timing of ovulation by exogenous and endogenous factors. In G. W. Potts and R. J. Wootton, eds., *Fish reproduction: strategies and tactics.* London: Academic Press.

Stearns, S. C. 1976. Life-history tactics: a review of the ideas. *Quarterly Review of Biology,* 51:3–47.

Steele, J. H. 1974. *The structure of marine ecosystems.* Cambridge, Mass.: Harvard University Press.

———— 1978. Some comments on plankton patches. In J. H. Steele, ed., *Spatial pattern in plankton communities.* NATO Conference Series 4, Marine Sciences, vol. 3. New York: Plenum Press.

Steele, J. H., and E. W. Henderson. 1984. Modeling long-term fluctuations in fish stock. *Science,* 224:985–987.

Stepien, W. P., Jr. 1976. Feeding of laboratory-reared larvae of the sea bream *Archosargus rhomboidalis* (Sparidae). *Mar. Biol.,* 38:1–16.

Stommel, H. 1963. Varieties of oceanographic experience. *Science,* 139:(3555):572–576.

———— 1965. Some thoughts about planning the Kuroshio survey. *Proc. Symp. on the Kuroshio.* Tokyo, October 29, 1963, Oceanogr. Soc. Japan and UNESCO.

Strasburg, D. W. 1959. An instance of natural mass mortality of larval frigate mackerel in the Hawaiian Islands. *J. Cons. int. Explor. Mer,* 24:255–263.

Sutcliffe, W. H., Jr., K. Drinkwater, and B. S. Muir. 1977. Correlations of fish catch and environmental factors in the Gulf of Maine. *J. Fish. Res. Board Can.,* 33:98–115.

Svardson, G. 1949. Natural selection and egg number in fish. *Rep. Inst. Freshwat. Res. Drottningholm,* 29:115–122.

Talbot, J. W. 1977. The dispersal of plaice eggs and larvae in the southern bight of the North Sea. *J. Cons int. Explor. Mer,* 37:221–248.

———— 1978. Changes in plaice larval dispersal in the last fifteen years. *Rapp. P.-v. Réun. Cons. int. Explor. Mer,* 172:114–123.

Tang, Q. 1985. Modification of the Ricker stock-recruitment model to account for environmentally induced variation in recruitment with particular reference to the blue crab fishery in Chesapeake Bay. *Fish. Res.,* 3:13–21.

Taylor, F. H. C. 1971. Variation in hatching success in Pacific herring *(Clupea pallasii)* eggs with water depth, temperature, salinity and egg mass thickness. *Rapp. P.-v. Réun. Cons. int. Explor. Mer,* 160:34–41.

Templeman, W., V. M. Hodder, and R. Wells. 1978. Age, growth, year-class strength, and mortality of the haddock, *Melanogrammus aeglefinus,* on the southern Grand Bank and their relation to the haddock fishery of this area. *ICNAF Res. Bull.,* 13:31–52.

Theilacker, G. H., and K. Dorsey. 1980. *Larval fish diversity, a summary of laboratory and field research.* Intergovernmental Oceanography Commission Workshop Report No. 28:105–142.

Theilacker, G. H., and M. F. McMaster. 1971. Mass culture of the rotifer *Brachionus plicatilis* and its evaluation as a food for larval anchovies. *Mar. Biol.,* 10:183–188.

Troadec, J.-P. 1983. Practices and prospects for fisheries development and management: The case of Northwest African fisheries. In B. J. Rothschild, ed., *Global fisheries: perspectives for the 1980s.* New York: Springer-Verlag.

Troadec, J.-P., W. G. Clark, and J. A. Gulland. 1980. A review of some pelagic fish stocks in other areas. *Rapp. P.-v. Réun. Cons. int. Explor. Mer,* 177:252–277.

Turner, M. E., R. J. Monroe, and H. J. Lucas, Jr. 1961. Generalized asymptotic regression and non-linear path analysis. *Biometrics,* 17:120–43.

Tyler, A. V., and R. S. Dunn. 1976. Ration, growth, and measures of somatic and organ condition in relation to meal frequency in winter flounder, *Pseudopleuronectes americanus,* with hypotheses regarding population homeostasis. *J. Fish. Res. Board Can.,* 33:63–75.

Uda, M. 1961. Fisheries oceanography in Japan, especially on the principles of fish distribution, concentration, dispersal and fluctuation. *Calif. Coop. Oceanic Fish. Invest. Rep.,* 8:25–31.

Ulanowicz, R. E., and T. Platt, eds. 1985. Ecosystem theory for biological oceanography. *Can. Bull. Fish. Aquat. Sci.,* 213:1–260.

Ulltang, Ø. 1984. The management of cod stocks with special reference to growth and recruitment overfishing and the question whether artificial propagation can help to solve management problems. In E. Dahl, D. S. Danielssen, E. Moksness, and P. Solemdal, eds., *The propagation of cod, Gadus morhua L.* Flødevigen rapportser 1, 1984:795–817.

Ursin, E. 1982. Stability and variability in the marine ecosystem. *Dana,* 2:51–67.

Vladykov, V. D. 1956. Fecundity of wild speckled trout *(Salvelinus fontinalis)* in Quebec lakes. *J. Fish. Res. Board Can.* 13:799–841.

Vlymen, W. J. 1977. A mathematical model of the relationship between larval anchovy *(Engraulis mordax)* growth, prey microdistribution, and larval behavior. *Env. Biol. Fish.,* 2:211–233.

Vogel, S. 1981. *Life in moving fluids.* Princeton, N. J.: Princeton University Press.

Volterra, V. 1926. Varizioni e fluttuazioni del numero l'individui in specie animali conviventi. *Mem. R. Acad. Naz. dei Lincei,* (ser. 6), 2:31–113.

Walford, L. A., and K. E. Mosher. 1943. *Studies on the Pacific pilchard.* 3. Determination of age of adults by scales and effect of environment on first year's growth as it bears on age determination. Spec. Sci. Rep. U.S. Fish. Wildl. Serv., Fisheries, No. 15:96–131.

Walters, C. J., and D. Ludwig. 1981. Effects of measurement errors on the assessment of stock-recruitment relationships. *Can. J. Fish. Aquat. Sci.,* 38:704–710.

Ware, D. M. 1975. Relation between egg size, growth, and natural mortality of larval fish. *J. Fish. Res. Board Can.,* 32:2503–512.

——— 1980. Bioenergetics of stock and recruitment. *Can. J. Fish. Aquat. Sci.,* 37:1012–24.

Ware, D. M., and T. C. Lambert. 1985. Early life history of Atlantic mackerel *(Scomber scombrus)* in the southern gulf of St. Lawrence. *Can. J. Fish. Aquat. Sci.,* 42: (in press).

Wiborg, Kr. Fr. 1976. Larval mortality in marine fishes and the critical period concept. *J. Cons. int. Explor. Mer,* 37:111.

Williams, R. 1984. An overview of secondary production in pelagic ecosystems. In M. J. R. Fasham, ed., *Flows of energy and materials in marine ecosystems.* NATO Conference Series IV, Marine Sciences; v. 13. New York: Plenum Press.

Woods, J. D. 1977. Parameterization of unresolved motions. In E. B. Kraus, ed., *Modelling and prediction of the upper layers of the ocean.* Oxford: Pergamon Press.

Woods, J. D., and W. Barkmann. 1986. The response of the upper ocean to solar heating I: The mixed layer. *Q. J. R. Meterol. Soc.,* 3: (in press).

Wootton, R. J. 1973. The effect of food ration on egg production in the female three-spined stickleback, *Gasterosteus aculeatus* L.

Wroblewski, J. S. 1984. Formulation of growth and mortality of larval northern anchovy in a turbulent feeding environment. *Mar. Ecol. Prog. Ser.,* 20: 13–22.

Wyatt, T. 1972. Some effects of food density on the growth and behavior of plaice larvae. *Mar. Biol.,* 14:210–216.

Wynne-Edwards, J. C. 1962. *Animal dispersion in relation to social behavior.* Edinburgh and New York: Hafner.

Yamomoto, T. 1961. Physiology of fertilization in fish eggs. *Int. Rev. Cytol.,* 12:361–405.

Zeeman, E. C. 1976. Catastrophe theory. *Scient. Am.,* 234:65–83.

Zweifel, J. R. 1973. A non-parametric approach to the estimation of relative change in fish population size from egg and larval surveys. *Rapp. P.-v. Réun. Cons. int. Explor. Mer,* 164:276–281.

Index